MW00843366

# Cathodic Protection

**Scrivener Publishing**
100 Cummings Center, Suite 541J
Beverly, MA 01915-6106

*Publishers at Scrivener*
Martin Scrivener (martin@scrivenerpublishing.com)
Phillip Carmical (pcarmical@scrivenerpublishing.com)

# Cathodic Protection

## Industrial Solutions for Protecting Against Corrosion

**Volkan Cicek**

WILEY

Co-published by John Wiley & Sons, Inc. Hoboken, New Jersey, and Scrivener Publishing LLC, Salem, Massachusetts.
Published simultaneously in Canada.



For general information on our other products and services or for technical support, please contact our Customer Care Department within the United States at (800) 762-2974, outside the United States at (317) 572-3993 or fax (317) 572-4002.

Wiley also publishes its books in a variety of electronic formats. Some content that appears in print may not be available in electronic formats. For more information about Wiley products, visit our web site at www.wiley.com.

For more information about Scrivener products please visit www.scrivenerpublishing.com.

Cover design by Kris Hackerott

***Library of Congress Cataloging-in-Publication Data:***

ISBN 978-1-118-29040-8

Printed in the United States of America

10 9 8 7 6 5 4 3 2 1

# Contents

# Acknowledgements

I dedicate this book to my lovely son Furkan Ali and my daughter Zehra Nur. I also acknowledge Ishik University for the willingness to support my studies.

# Preface

In this book, cathodic protection as a corrosion prevention technique is detailed along with its underlying scientific background as it relates to corrosion chemistry, corrosion engineering, physical chemistry, and chemical engineering in general. In addition to the theoretical framework of the phenomenon, industrial practices are exemplified, and frequently encountered associated problems along with their solutions are described. Therefore, it is my wish that the reader will find this book a didactic one, and will attain a comprehensive knowledge of the subject.

Cathodic protection practice started in the 1930s. It was first implemented in petroleum pipelines, then in piers, ports, ships, water and petroleum storage tanks, reinforced concretes, etc., mainly in underground, water, and underwater systems. Cathodic protection is the process of converting the anodic metal into a cathode, preventing the anodic currents by externally providing the electrons that the cathodic reaction needs. This way, while the externally provided electrons stop the anodic reaction and thus the dissolution of the metal, they increase the rate of the cathodic reaction. Alternatively, the principle of cathodic protection technique is to establish potential conditions such that metal remains in the immunity zone as described in the pH-potential diagrams. There are two methods of changing metal's potential in the negative direction by providing an external current: sacrificial anode cathodic protection and impressed current cathodic protection.

In sacrificial anode cathodic protection, the host metal is protected by connecting a more active metal, forming a galvanic cell, where the host cell is the cathode. For the current to flow through this new sacrificial anode cathodic protection cell, there must be sufficient potential difference between the anode and the cathode to overcome the circuit resistance. Current withdrawn from the galvanic anode depends on the anode's open circuit potential and the circuit resistance. In this method, galvanic anode corrodes instead of the metal to be protected; thus, it has a limited service life. In sacrificial anode cathodic protection systems, the quantity and size of the anode can only be determined based on the minimum current intensity required for cathodic protection. In brief, sacrificial anodes' corrosion potentials must be sufficiently negative, their anodic capacity and anodic efficiencies must be high, and they should be continuously active and not passivated.

In cathodic protection systems, half of the initial establishment expense is spent for anodes; thus, it is important for the anodes to be economical. It is important that the current withdrawn from the anode is as high as possible, and anode resistance should not increase over time. Hence, mass loss per current withdrawn (A.year) must be as small as possible.

In impressed current cathodic protection, the potential of the metal that undergoes corrosion is changed, converting it to a cathode by connecting it to another system that is protected by an electrode made of a noble metal as the anode making up the negative pole of the corrosion cell. In practice, a direct current is applied, the current resource's negative pole is connected to the metal, and the positive pole is connected to the anode. Intensity of the applied current depends on the surface area of the metal to be protected and the corrosivity of the environment. In this method, since the energy or current is provided externally, the reference anode does not corrode instead of the metal to be protected; however, since every metal dissolves more or less with an applied potential, even the most durable anode to be chosen has a limited service life, although it can be very long. Parameters determining the current magnitude are the environment's resistivity, pH, dissolved $O_2$, etc. Therefore, instead of current adjustment, controlling the potential is usually the preferred method. In impressed current cathodic protection, for anodes such as Pt and Ti, issues such as location plan of anodes, connections, and potential losses along the connections must be considered. Additionally, locations where potentials will be measured must be carefully selected for successful monitoring of the system.

In the case of heterogeneous distribution of the current as well as overprotection, hydrogen gas evolves at the cathode, sometimes leading to hydrogen embrittlement, due to hydrogen atoms diffusing into the metal. If the applied current becomes a stray current, for instance, due to the anode bed being placed close to a railway, then cathodic protection may become ineffective. Circuit potential is usually low in sacrificial anode cathodic protection systems; thus, they cannot be applied in grounds with high resistivity values unless a galvanic anode with a high potential is used. Regularly, they are applied in grounds with resistivity up to 5000 ohm.cm. On the other hand, high resistivity values of the ground do not cause a problem for impressed current cathodic protection systems. By reducing the resistance of the anode bed, for instance, current can be adjusted to the desired intensities.

Sacrificial anode cathodic protection is very easy to install and apply. Additionally, if more current is needed later on, more anodes can be installed. On the other hand, for impressed current cathodic protection systems, the intensity of the current depends on the capacity of the transformer/rectifier unit; also, the resistance of the anode bed cannot be reduced during operation. In sacrificial anode cathodic protection systems, it is not possible to manually adjust the intensity of the current withdrawn from the anodes. Galvanic anodes adjust the needed current level automatically. When more current is needed, the potential of the structure decreases, increasing the difference in potential between the anode and the cathode, leading to more current being withdrawn from the anode. In impressed current cathodic protection systems, changes in the amount of the applied current should be done manually, or the system has to be set to automatically perform the needed changes so that the protection potential does not fall below a certain limit. In sacrificial anode cathodic protection systems, metal dissolution due to application of high potentials in the surroundings of the anode is not observed, while it occurs in impressed current cathodic protection systems. Interference effects are negligible in sacrificial anode cathodic protection systems, since their anode-ground potentials are low, while interference effects may occur in impressed current cathodic protection systems around the anodic beds and at the intersections of pipelines that are cathodically protected with those of that are not.

Unit cost of the current provided by sacrificial anode cathodic protection is higher than it is by the impressed current cathodic protection; thus, sacrificial anode cathodic protection is not preferred in pipelines that require high currents. Initial establishment costs of the impressed current cathodic protection system are higher than that of sacrificial anode cathodic protection system. However, sacrificial anode cathodic protection is the only applicable method in areas where it is not possible to generate electricity power. In sacrificial anode cathodic protection, there is no need for an external current source, since galvanic anodes are the source of the required currents.

# 1

# Corrosion of Materials

Corrosion comes from Latin word "corrodere." Plato talked about corrosion first in his lifetime (B.C. 427–347), defining rust as a component similar to soil separated from the metal. Almost 2000 years later, Georgius Agricola gave a similar definition of rust in his book entitled *Mineralogy*, stating that rust is a secretion of metal and can be protected via a coating of tar. The corrosion process is mentioned again in 1667 in a French-German translation, and in 1836 in another translation done by Sir Humphrey Davy from French to English, where cathodic protection of metallic iron in seawater is mentioned. Around the same time, Michael Faraday developed the formulas defining generation of an electrical current due to electrochemical reactions.

To one degree or another, most materials experience some type of interaction with a large number of diverse environments. Often, such interactions impair a material's usefulness as a result of the deterioration of its mechanical properties, e.g., ductility, strength, other physical properties, and appearance.

Deteriorative mechanisms are different for three material types, which are ceramics, polymers, and metals. Ceramic materials are relatively resistant to deterioration, which usually occurs at elevated temperatures or in extreme environments; that process is also frequently called "corrosion." In the case of polymers, mechanisms and consequences differ from those for metals and ceramics, and the term "degradation" is most frequently used. Polymers may dissolve when exposed to liquid solvent, or they may absorb the solvent and swell. Additionally, electromagnetic radiation, e.g., primarily ultraviolet and heat, may cause alterations in their molecular structures. Finally, in metals, there is actual material loss, either by dissolution or corrosion, or by the formation of a film or nonmetallic scales by oxidation; this process is entitled "corrosion" as well.

## 1.1 Deterioration or Corrosion of Ceramic Materials

Ceramic materials, which are sort of intermediate compounds between metallic and nonmetallic elements, may be thought of as having already been corroded. Thus, they are exceedingly immune to corrosion by almost all environments, especially at room temperature, which is why they are frequently utilized. Glass is often used to contain liquids for this reason.

Corrosion of ceramic materials generally involves simple chemical dissolution, in contrast to the electrochemical processes found in metals. Refractory ceramics must not only withstand high temperatures and provide thermal insulation, but in many instances, must also resist high temperature attack by molten metals, salts, slags, and glasses. Some of the more useful new technology schemes for converting energy from one form to another require relatively high temperatures, corrosive atmospheres, and pressures above the ambient. Ceramic materials are much better suited to withstand most of these environments for reasonable time periods than are metals.

## 1.2 Degradation or Deterioration of Polymers

Polymeric materials deteriorate by noncorrosive processes. Upon exposure to liquids, they may experience degradation by swelling or dissolution. With swelling, solute molecules actually fit into the molecular structure. Scission, or the severance of molecular chain bonds, may be induced by radiation, chemical reactions, or heat. This results in a reduction of molecular weight and a deterioration of the physical and chemical properties of the polymer.

Polymeric materials also experience deterioration by means of environmental interactions. However, an undesirable interaction is specified as degradation, rather than corrosion, because the processes are basically dissimilar. Whereas most metallic corrosion reactions are electrochemical, by contrast, polymeric degradation is physiochemical; that is, it involves physical as well as chemical phenomena. Furthermore, a wide variety of reactions and adverse consequences are possible for polymer degradation. Covalent bond rupture, as a result of heat energy, chemical reactions, and radiation is also possible, ordinarily with an attendant reduction in mechanical integrity. It should also be mentioned that because of the chemical complexity of polymers, their degradation mechanisms are not well understood.

Polyethylene (PE), for instance, suffers an impairment of its mechanical properties by becoming brittle when exposed to high temperatures in an oxygen atmosphere. In another example, the utility of polyvinylchloride (PVC) may be limited because it is colored when exposed to high temperatures, even though such environments do not affect its mechanical characteristics.

## 1.3 Corrosion or Deterioration of Metals

Among the three types of materials that deteriorate, "corrosion" is usually referred to the destructive and unintentional attack of a metal, which is an electrochemical process and ordinarily begins at the surface. The corrosion of a metal or an alloy can be determined either by direct determination of change in weight in a given environment or via changes in physical, electrical, or electrochemical properties with time.

In nature, most metals are found in a chemically combined state known as an ore. All of the metals (except noble metals such as gold, platinum, and silver) exist in nature in the form of their oxides, hydroxides, carbonates, silicates, sulfides, sulfates, etc., which are thermodynamically more stable low-energy states. The metals are extracted from these ores after supplying a large amount of energy, obtaining pure metals in their elemental forms. Thermodynamically, as a result of this metallurgical process, metals attain higher energy levels, their entropies are reduced, and they become relatively more unstable, which is the driving force behind corrosion. It is a natural tendency to go back to their oxidized states of lower energies, to their combined states, by recombining with the elements present in the environment, resulting in a net decrease in free energy.

Since the main theme of the book is cathodic protection, which is a preventive measure against corrosion of metals, the remainder of the chapter will focus on the associated corrosion processes of widely used metals before going into details of cathodic protection. One shall have an idea about the corrosion process so that a more comprehensive understanding of cathodic protection process is possible.

Therefore, first, commonly used metals will be reviewed in terms of their corrosion tendencies, beginning with iron and steel, which are the most commonly used structural metals, and thus the most commonly protected ones with cathodic protection.

### 1.3.1 Iron, Steel and Stainless Steels

Iron and steel makes up 90% of all of the metals produced on earth, with most of it being low carbon steel. Low carbon steel is the most convenient metal to be used for machinery and equipment production, due to its mechanical properties and low cost. An example is the pressurized containers made of carbon steel that has 0.1% to 0.35% carbon. Carbon steel costs one-third as much as lead and zinc, one-sixth as much as aluminum and copper, and one-twentieth as much as nickel alloys. However, the biggest disadvantage of carbon steel is its low resistance to corrosion.

The most common mineral of iron in nature is hematite ($Fe_2O_3$), which is reacted with coke dust in high temperature ovens to obtain metallic iron. 1 ton of coke dust is used to produce 1 ton of iron. The naturally occurring reverse reaction, which is corrosion of iron back to its mineral form, also consists of similar products to hematite such as iron oxides and hydroxides. Energy released during the corrosion reaction is the driving factor for the reaction to be a spontaneous reaction; however, in some cases, even if the free Gibbs energy of the reaction is negative, due to a very slow reaction rate, corrosion can be considered as a negligible reaction, such as in the cases of passivation and formation of naturally protective oxide films.

The anodic reactions during the corrosion of iron under different conditions are the same, and it is clearly the oxidation of iron producing $Fe^{2+}$ cations and electrons. However, the cathodic reaction depends on the conditions to which iron is exposed. For example, when no or little oxygen is present, like the iron pipes buried in soil, reduction of $H^+$ and water occurs, leading to the evolution of hydrogen gas and hydroxide ions. Since iron (II) hydroxide is less soluble, it is deposited on the metal surface and inhibits further oxidation of iron to some extent.

$$Fe \longrightarrow Fe^{2+} + 2e^- \quad \text{(Eq. 1)}$$

$$2H_2O + 2e^- \longrightarrow 2OH^- + H_2 \quad \text{(Eq. 2)}$$

$$Fe + 2H_2O \longrightarrow \underbrace{Fe^{2+} + 2OH^-}_{Fe(OH)_2} + H_2 \quad \text{(Eq. 3)}$$

Thus, corrosion of iron in the absence of oxygen is slow. The product, iron (II) hydroxide, is further oxidized to magnetic iron oxide or magnetite that is $Fe_3O_4$, which is a mixed oxide of $Fe_2O_3$ and FeO. Therefore, an iron object buried in soil corrodes due to the formation of black magnetite on its surface.

$$6Fe(OH)_2 + O_2 \longrightarrow 2Fe_3O_4.H_2O + 4H_2O \quad \text{(Eq. 4)}$$

$$Fe_3O_4.H_2O \longrightarrow \underbrace{H_2O + Fe_3O_4}_{\text{black magnetite}} \quad \text{(Eq. 5)}$$

If oxygen and water are present, the cathodic reactions of corrosion are different. In this case, the corrosion occurs about 100 times faster than in the absence of oxygen. The reactions involved are:

$$2 \times (Fe \longrightarrow Fe^{2+} + 2e^-) \quad \text{(Eq. 6)}$$

$$O_2 + 2H_2O + 4e^- \longrightarrow 4OH^- \quad \text{(Eq. 7)}$$

$$2Fe + O_2 + 2H_2O \longrightarrow 2Fe(OH)_2 \quad \text{(Eq. 8)}$$

As oxygen is freely available, the product, iron (II) hydroxide, further reacts with oxygen to give red-brown iron (II) oxide:

$$4Fe(OH)_2 + O_2 \longrightarrow \underbrace{2Fe_2O_3.H_2O}_{\text{red brown}} + 2H_2O \quad \text{(Eq. 9)}$$

The red brown rust is the most familiar form of rust, since it is commonly visible on iron objects, cars, and sometimes in tap water. The process of rusting is increased due to chlorides carried by winds from the sea, since chloride can diffuse into metal oxide coatings and form metal chlorides, which are more soluble than oxides or hydroxides. The metal chloride so formed leaches back to the surface, and thus opens a path for further attack of iron by oxygen and water.

Presence of pollutants in the air affects the rate of corrosion. $SO_2$ is a notorious air pollutant, usually formed by the combustion of coal in power plants or in homes. The solubility of $SO_2$ in water is about 1000 times greater than $O_2$ and is the reason for the formation of sulfuric acid and so-called acid rain, leading to following corrosion reactions:

$$Fe + SO_2 + O_2 \longrightarrow FeSO_4 \quad \text{(Eq. 10)}$$

$$4FeSO_4 + O_2 + 6H_2O \longrightarrow 2FeO_3.H_2O + 4H_2SO_4 \quad \text{(Eq. 11)}$$

$$4H_2SO_4 + 4Fe + 2O_2 \longrightarrow 4FeSO_4 + 4H_2O \quad \text{(Eq. 12)}$$

The sulfuric acid formed in these reactions is difficult to remove, which is why, even after cleaning the iron object carefully, corrosion continues as long as sulfates are present in the medium. However, the effect of sulfate ions on iron corrosion in chloride solutions was found to be weak up to pH 5.5, while above pH 5.5, sulfate ions act as weak inhibitors. Iron's anodic reactions in sulfate solution within pH range of 0 to 6 are as follows:

$$Fe + H_2O \longleftrightarrow Fe(H_2O)_{ads} \quad \text{(Eq. 13)}$$

$$Fe(H_2O)_{ads} \longleftrightarrow Fe(OH^-)_{ads} + H^+ \quad \text{(Eq. 14)}$$

$$Fe(OH^-)_{ads} \longleftrightarrow Fe(OH)_{ads} + e^- \quad \text{(Eq. 15)}$$

$$Fe(OH)_{ads} \longleftrightarrow FeOH^+ + e^- \quad \text{(Eq. 16)}$$

$$FeOH^+ + H^+ \longleftrightarrow Fe^{2+} + H_2O \quad \text{(Eq. 17)}$$

Since pure iron is relatively softer, it is alloyed with elements such as Cr, Ni, Mn, Co, Si, Al, Ti, V, W, and Zi to make it harder and stronger. Steel is such an alloy with elements C, Mn, Si, S, and P.

Stainless steels have certain alloying elements in sufficient amounts in their composition so a passive layer can form on their surface, preventing corrosion and increasing its mechanical properties. These elements are primarily chromium of amounts less than 10.5% and carbon of amounts less than 1.2%. Stainless steels are mostly used in chemistry.

The most common stainless steel is *austenitic steel*, which is not magnetic and makes up more than 65% of all stainless steels used in the world, has less than 0.1% carbon in content, and is primarily made up of iron, chromium and nickel as alloying elements. Other commonly used stainless steels are *ferritic steel,* which has magnetic characteristics and is mainly iron and chromium with less than 0.1% carbon, *martensitic steel,* which can be hardened, is magnetic, and is mainly iron and chromium with more than 0.1% carbon, and *double phased or duplex steel,* which is magnetic, is made up of iron, chromium, and nickel, and is basically a combination of austenitic and ferritic steel.

Most of the stainless steels are exposed heavily to pitting corrosion and stress corrosion cracking in seawater that has abundant chlorides and oxygen. For stainless steels to passivate, the chromium percentage in the alloy must be more than 12%; however, due to precipitation in the form of $Cr_{23}C_6$ with the carbon in steel, a higher percentage may be needed. Another alloying element other than carbon, chromium, and nickel is molybdenum, which is known as ferrite maker and is added to austenitic steels in the amount of 2% to 3%, increasing the resistance to pitting corrosion in presence of chlorides. However, addition of molybdenum also reduces the corrosion resistance of 18Cr-10Ni stainless steel in hot nitric acid. Titanium 321 and Niobium 347 can be added to austenitic steels to reduce their sensitivity against some types of corrosion. Additionally, copper can be added to increase corrosion resistance against oxidizing acids, acidic environments in general, and chlorides. Selenium and sulfur increase the mechanical properties of stainless steels such as malleability, while silicon reduces the stainless steels' tendency to oxidize at high temperatures.

*Martensitic stainless steels* have 12% to 20% chromium and low carbon. They can be hardened via thermal treatment. Their corrosion resistance is more than mild steel but less than austenitic steels. They can be used safely in mildly corrosive environments, such as in the atmosphere or in fresh waters, and in temperatures up to 650°C.

*Ferritic stainless steels* have 15% to 30% chromium in their composition, more than martensitic steels have, which is why they are more resistant to corrosion. They can be used in chemical equipments, storage tanks, and kitchenware.

*Austenitic stainless steels* are alloys of chromium and nickel. 300 series austenitic steels, for instance, have 16% to 26% chromium and 7% to 22% nickel in their composition. They are easily shaped, are highly resistant to corrosion, and can be welded such as widely used AISI 304 18–8 steel. 200 series austenitic steels have manganese and nitrogen in their composition as well. They are mechanically superior compared to 300 series, but inferior in terms of their corrosion resistance.

*Double phased or duplex stainless steels* are also alloys of chromium and nickel, but with one phase of austenitic steels and another phase of ferritic steels, giving them a composition of 28% chromium and 6% nickel. In terms of their mechanical and corrosion resistance properties, they take place in between austenitic steels and ferritic steels. They are very resistant to stress corrosion and intergranular corrosion.

*Stainless steels that are hardened via precipitation* are special type of steels that have high strength/weight ratio and high corrosion resistance; thus, they are used in aircraft and space industries. They are produced in three types: martensitic, half austenitic, and austenitic.

### 1.3.2 Aluminum and Its Alloys

Aluminum is extensively used because it has a low density that is 2.7 g/cm$^3$, it has high thermal and electrical conductivity, its alloys made with thermal operations have high mechanical strength, and it has high corrosion resistivity compared with other pure metals. Normally, aluminum is more active than all metals but alkaline and earth alkaline metals in electrochemical series, and thus should have acted as anodic towards all other elements of the periodic table; however, due to the oxide layer that passivates its surface, it is quite resistant to corrosion. It is very resistant to water, organic acids, and some oxidizing acids. Therefore, it is frequently used in reaction containers, machinery equipment, and chemical batteries, e.g., aluminum tanks are used to carry acetic acid.

The $Al_2O_3$ layer that protects aluminum from corrosion forms very quickly due to the high reactive nature of aluminum, and this layer can also be produced via electrical current in laboratory conditions. Chatalov first studied the aluminum corrosion based on pH in 1952, while Pourbaix and colleagues found out that corrosion rate logarithmically depends on pH, and that the least corrosion takes place when pH is 6, because aluminum hydrates that form as corrosion products have the least solubility amounts at this level. Binger and Marstiller found the same logarithmic relation for pH between 7 and 10. Vujicic and Lovrecek claimed that corrosion rate depending on pH is 50% more than that suggested by Chatalov. Tabrizi, Lyon, and colleagues found that with increasing pH from 8 to 11, corrosion rate increases, while it slows down at pH 11 and increases again at pH 12. They also found that with increasing temperature, corrosion rate also increases.

As a result of these studies, it is generally accepted that aluminum is passive in the pH range of 4 to 9, and forms a non-permeable and insulating oxide film. Aluminum metal surface has zero charge at pH 9.1. Aluminum corrodes or dissolves when the pH is out of its passivity range; however, it dissolves less in acidic mediums than it does in basic media. In alkaline environments, aluminum and alloys are easily corroded, especially for pH values over 10. Therefore, in the case of cathodic protection application, overprotection must be avoided, since it will lead to an increase in pH. Damaging of protective $Al_2O_3$ layer occurs based on the following reaction in basic medium:

$$Al_2O_3 + 2OH^- + 3H_2O \longleftrightarrow 2Al(OH)_4^- \qquad \text{(Eq. 18)}$$

In $NH_3$ solutions over pH 11.5, $NH_3$ dominates its conjugate acid $NH_4^+$ in the buffer system, and resistance of the system towards corrosion increases because $NH_3$ is a stronger ligand than $OH^-$; thus, $OH^-$ cannot bind to aluminum and dissolve it away. Therefore, dissolution slows down and corrosion current lessens: opposite to what is observed in KOH solutions in the same pH range.

When studying the effect of sulfate ions on aluminum corrosion, aluminum corrosion in less concentrated $Na_2SO_4$ and $H_2SO_4$ solutions with pH values of 1.5 were found as $10^{-4}$ mA.cm$^{-2}$ and 1.24 $10^{-4}$ mA.cm$^{-2}$; thus, aluminum is stable in less concentrated sulfuric acid solutions, but not in concentrated solutions. When $Na_2SO_4$ is added to the system or in the presence of $SO_2$ in atmospheric conditions, corrosion of aluminum increases substantially at pH 12 due to a large increase in the conductivity of the solution. The opposite occurs for corrosion in KOH, since sulfate ions competitively adsorb at the aluminum surface, with $OH^-$ ions lessening the corrosion up to 50%. In acidic medium, in HCl solution, aluminum dissolves as follows:

$$Al + Cl^- \longleftrightarrow AlCl^-_{ads} \qquad \text{(Eq. 19)}$$

$$AlCl^-_{ads} + Cl^- \longrightarrow AlCl^+_2 + 3e^- \text{ (slow)} \qquad \text{(Eq. 20)}$$

When organic inhibitors are used to prevent aluminum corrosion, protonated organic inhibitors adsorb at the metal surface through $AlCl^-_{ads}$ preventing $AlCl^-_{ads}$ from oxidizing into $AlCl^+_2$. Protonated organic inhibitors may also stabilize chlorides, thus preventing chlorides from reacting. While hydrogen gas evolution takes place at the cathode:

$$H^+ + e^- \longrightarrow H_{ads} \text{ (fast)} \qquad \text{(Eq. 21)}$$

$$H^+ + H_{ads} \longrightarrow H_2 \text{ (slow)} \qquad \text{(Eq. 22)}$$

Protonated organic inhibitors may competitively adsorb on the metal surface with respect to hydrogen, and thus also may prevent cathodic hydrogen evolution.

Aluminum alloys that have high aluminum content are susceptible to stress corrosion; thus, they are coated with pure aluminum, making it Alclad aluminum. There are many such Alclad aluminum alloys of high strength containing Mg and Si. Halogenated organic compounds may damage aluminum materials by reacting with them over time. Aluminum and its alloys have become very valuable due to its wide use in different areas of the industry. Its value in the London Metal Exchange has increased to $3380 per ton in 2008. Due to this increase in aluminum prices, and thus due to the increase in the costs of employing aluminum components, corrosion prevention of aluminum and aluminum alloys became even more important.

One of the corrosion prevention methods is anodic oxidation, or anodizing the aluminum surface to develop the naturally occurring aluminum oxide layer on the surface of the aluminum, making the naturally formed 25 A° layer thicker. An artificially developed aluminum oxide layer has levels of corrosion resistance depending on the conditions of the anodizing process such as the electrolyte type, applied potential, application duration, application temperature, etc. Most commonly used anodizing electrolytes or solutions are solutions of sulfuric, boric, oxalic, phosphoric or chromic acids. Among these, chromic acid forms the protective aluminum oxide layers automatically, but it is toxic, oxalic acid decomposes at high temperatures since it is an organic acid, phosphoric acid requires high anodizing potentials increasing the costs, and sulfuric and boric acids seem to be more convenient in general applications since they are not toxic, they are economical, and they are easily obtainable.

### 1.3.3 Magnesium and Its Alloys

Magnesium alloys are used in automobile and other industries because they are light, but their low corrosion resistance limits their use. They are also widely used as anodes in the cathodic protection systems. Magnesium oxide film that is formed on magnesium surface is easily affected by the chlorides and acids; thus, magnesium surfaces must be treated with passivating ions such as fluorides, phosphates, and chromates to form a strong protective film. Then they can be used in aircraft and automobile industries. Along with chromate coatings, coatings of tungstates, molibdates, silicates, borates, and lanthanides are also used to protect magnesium alloys. These metals form lowly soluble compounds with magnesium cations; however, some among them, especially chromates, are very toxic. Even coated magnesium alloys would still be susceptible to stress corrosion cracking in presence of chlorides. Due to dissolution of magnesium hydroxide, which is the corrosion product of magnesium, pH increases, making it difficult to keep pH neutral.

### 1.3.4 Copper and Its Alloys

Copper and its alloys are used extensively in industry, especially in cooling and heating systems, due to their very high electrical and thermal conductivity and appropriate mechanical properties making them ideal heat transfer materials. In water heating systems, they are used as pure copper or as alloys of zinc, tin, or nickel. Brass is an alloy consisting of ~70% Cu and ~30% Zn, and copper-tin alloy is called bronze. Both brass and bronze could be improved with the addition of silicon and beryllium. Beryllium-copper alloys can provide mechanical strength up to 1300 MPa and they are safe to use along with explosives. Copper alloys are susceptible to stress corrosion in the presence of ammines and ammonia. Additionally, brass is susceptible to selective corrosion, called dezinfication, unless it is inhibited with metals of group 5.

Copper's thermal conductivity is reduced due to the formation of a layer of corrosion products on the metal surface. Pickling acidic solutions are used to clean the surface from corrosion precipitates. In sulfuric acid, for instance, copper anodically dissolves as follows:

$$2Cu + H_2O \longrightarrow Cu_2O + 2H^+ + 2e^- \quad \text{(Eq. 23)}$$

$$Cu_2O + 2H^+ \longrightarrow 2Cu^{+2} + H_2O + 2e^- \quad \text{(Eq. 24)}$$

The pH values that are either neutral or near neutral lead to the formation of copper oxides and hydroxides on the copper surface. In the presence of chlorides, copper dissolves as follows:

$$Cu + Cl^- \longleftrightarrow CuCl + e^- \quad \text{(Eq. 25)}$$

$$CuCl + Cl^- \longrightarrow CuCl_2^- \quad \text{(Eq. 26)}$$

Meanwhile, copper oxide and hydroxides can also form in presence of chlorides. Cathodic reaction is the reduction reaction of oxygen:

$$O_2 + 2H_2O + 4e^- \longrightarrow 4OH^- \qquad \text{(Eq. 27)}$$

### 1.3.5 Nickel and Its Alloys

Nickel alloys are the best materials for alkali environments. Their alloys with stainless steels and others are used at high temperature applications. Nickel alloys are also resistant to chemical effects. In the case of hydrogen fluoride exposure, monel alloys, and in other cases, hastelloy, chlorimet, and inconel alloys are used. Nickel alloys are suitable for welding as well, unless there is lead or sulfur in the alloy, which would cause cracking when welded.

### 1.3.6 Titanium and Its Alloys

Titanium is the most appropriate metal for the aircraft and chemistry industries. Similar to aluminum oxide, the titanium oxide ($TiO_2$) layer that naturally forms on the surface of titanium is very protective against corrosion. Titanium is light, with a density of 4.5 g/cm$^3$; thus it covers 80% more volume when compared to stainless steel of the same weight. Due to its superior mechanical properties, very thin titanium pipes can be as strong as pipes made of other materials, although its low elasticity may cause problems, especially if there is vibration. Since titanium is expensive, it can be economical only if used in equipment and materials that are planned to have long service lives. Impurities such as nitrogen and oxygen in the titanium's crystalline structure reduce its resistance to corrosion, causing it to break dangerously, especially at higher temperatures.

### 1.3.7 Lead and Its Alloys

Lead is used in sulfuric acid industry due to being resistant to chemical effects; however, it is soft, and it has a density of 11.3 g/cm$^3$, making it heavy. Its mechanical strength is increased by alloying with antimony when used in a container. Its alloys with tin, tellurium, and calcium also have very good corrosion resistance characteristics. However, when welded, the produced vapors are hazardous, and must be ventilated well. Lead's corrosion products are toxic as well. Since lead is anodic compared to copper, water containers and pipes made of copper must not be welded with lead based welding; instead, silver welding can be used.

### 1.3.8 Corrosion of Composite Alloys (Tin Can Example)

Corrosion of tin cans is a good example of the corrosion of composite alloys in different mediums when used for different purposes. Cans are 0.11 to 0.3 mm thick low carbon steels coated with tin on each side, giving it a bright appearance. They can be easily shaped, welded, and soldered. Tin cans have 99% steel by weight; the tin layer is less thick than 0.0025 mm, usually between 0.00038 mm and 0.0015 mm. Since it is very thin, the amount of tin in tinned cans is given as $g/m^2$ and not as thickness, which is between 2.8 g and 11.2 g per square meter, which is doubled, considering two sides of the can. Electrically chromated steel or tin-free steel cans, which were developed first in Japan in 1965 due to tin being expensive, can be produced as bright- or dull-colored. They have to be used with a lacquer layer, to which they adhere very well, but they cannot be soldered. Without lacquer, severe corrosion occurs, leading to inflation of the can due to $H_2$ evolution. Due to well combination with lacquer layer, corrosion under the lacquer layer is prevented, as well as color changes due to sulfur. Chromated cans have less chromium compared to tins in the form of metallic chromium and chromium oxide, with 50 to 100 mg per $m^2$, compared to 2.8 g to 11.2 g in tin cans. Chromated cans are more resistant to alkaline environment than the acidic environment. Since they cannot be soldered, they have to be welded or affixed. One advantage of welding over soldering is that there is no lead and tin involvement that can diffuse into the food; also, welded cans are stronger than soldered ones. Chromated cans are better than tin cans in terms of being painted and lacquer applicability, as well as thermal durability and resistance against yellow discoloring, while worse in terms of being soldered and welded, and the same in terms of resistance to corrosion and malleability.

As a result of corrosion in tin cans, the quality of canned food decreases, and it becomes unhealthy due to diffusion of elements such as tin, iron, aluminum, lead, and cadmium into the food. At severe corrosion conditions, cans are punctured, diminishing the microbiological durability of the canned food, resulting in product loss. $H_2$ gas produced due to the corrosion inside the can causes the can to expand. Tin cans have 5 layers; the inner layer is the steel followed by tin-iron ($FeSn_2$) layer, then tin layer with the tin oxide layer above and lubricant layer on top, and with the optional lacquer layer as the sixth layer overall. Corrosion increases with the increasing ratio of sulfur and phosphorus in steel's composition, and decreases with increasing copper ratio. In any case, sulfur percentage should not be more than 0.4% not to lead to accelerated corrosion. Anodic corrosion reaction is:

$$Sn \longrightarrow Sn^{2+} + 2e^- \qquad \text{(Eq. 28)}$$

Cathodic corrosion reaction is:

$$2H^+ + 2e^- \longrightarrow H_2 \qquad \text{(Eq. 29)}$$

Molecules and ions such as $H^+$ consume electrons, preventing polarization until the electrode is fully covered with $H_2$. Such molecules and ions preventing polarization are called depolarizators. Oxygen present in the environment also acts as a depolarizator by reacting with the $H_2$ that covers cathode continuing the corrosion process:

$$2H^+ + 2e^- \longrightarrow 2H \qquad \text{(Eq. 30)}$$

Instead of forming $H_2$, hydrogen atoms react with oxygen producing water:

$$2H + \tfrac{1}{2} O_2 \longrightarrow H_2O \qquad \text{(Eq. 31)}$$

Thus, if oxygen is present in the environment, corrosion progresses continuously, resulting in iron being exposed and acting as anode, making tin the cathode in return; this leads to pitting corrosion, and the can to be punctured eventually. Hence, the degassing process is very important to prevent this type of corrosion, especially for food that is corrosive, such as fruits, which have to be canned under strong vacuum. Under normal conditions, oxygen gas that remains inside the cans is consumed in a few days. After the hydrogen gas produced by the reduced ions fully cover the tin cathode, tin becomes anode, and the resulting iron surface acts as the cathode on which hydrogen gas formation continues on sites where the iron surface is exposed. Since the exposed area of the steel is very small compared to the area of the tin surface, corrosion reaction slows down, and is said to be under cathodic control. Thus, inflation or expansion of a tin can due to hydrogen gas formation can be observed only after several months. Strongly corrosive foods are mostly fruits such as strawberries and other berries, cherries and sour cherries, plums, apple juice and cider, pickles, etc. Medium corrosive foods are peaches, apricots, figs, pears, grapefruit, etc., and weakly corrosive food are mostly vegetables such as peas, green beans, tomatoes, meat, fish, etc. Generally, compounds in fruits that cause corrosion are organic acids, anthocyanins, flavanols, catechins, hydroxymethylfurfurals, sulfur containing compounds, and oxygen. In vegetables, they are amino acids and proteins containing sulfur, oxygen, chlorides, nitrates, oxalic acid, ascorbic acid, and products of pectin decomposition. Storage conditions are also very important on corrosion. Every 10 degree increase in storage conditions doubles the corrosion that takes place. Thus, a can that has a shelf life of 1 year at 20°C would have 6 months of shelf life at 30°C and 3 months at 40°C.

Issue known as sulfur blackening in cans is due to the formation of metal sulfur compound at the surface of the can. Mainly, sulfur blackening is due to a mixture of tin sulfides and tin oxide. It occurs in canned food that has pH over 5, and thus never occurs in canned fruits. Sources of sulfur are amino acids such as methionine, cystine, cysteine, and peptides such as glutathione formed by the decomposition of the food, while the source of the metal is the tin layer. Compounds associated with sulfur blackening are SnS and $SnS_2$, which are black in color, and $Sn_2S_3$, which is red in color. Sulfur blackening mostly occurs in cans storing protein-rich food such as meat, fish, peas, and fava beans. Sulfur containing amino acids in these foods thermally decompose, producing $H_2S$ or other products containing thiol (-SH) group.

$$RC(O)SR + H_2O \longrightarrow RCOOH + RSH \quad \text{(Eq. 32)}$$

then thiol bond is hydrolyzed:

$$RSH + H_2O \longrightarrow ROH + H_2S \quad \text{(Eq. 33)}$$

then $H_2S$ reacts with the metal:

$$Sn + H_2S \longrightarrow SnS + H_2 \text{ or} \quad \text{(Eq. 34)}$$

$$Sn + 2H_2S \longrightarrow SnS_2 + 2H_2 \quad \text{(Eq. 35)}$$

$H_2S$ formation begins over 70°C and is completed mostly during the sterilization stage of the can, since the permeability of lacquer layer increases fivefold during sterilization. Metal sulfide formation rate increases over pH 6.15 and stops after 6 months of storage.

In general, corrosion prevention in cans can be done in the following ways: using appropriate tin and lacquer layer thickness based on the type of food canned; physically not scribing and damaging the surface; removing the oxygen present inside the can at the top; not having nitrate in the filling solution; using corrosive materials such as acids and salts in lesser concentrations; and using a lacquer layer.

# 2

# Factors Influencing Corrosion

Corrosion of a metal surface mainly depends on nature of metal and the nature of the corroding environment.

## 2.1 Nature of the Metal

### 2.1.1 Position in Galvanic Series

When two metals are in electrical contact, the more active metal higher up in the galvanic series that has the greater oxidation potential constitutes the anode in the presence of an electrolyte, and suffers corrosion. The rate and severity of corrosion depend upon the difference in their positions in the galvanic series. The greater the difference, the faster is the corrosion of anodic metal.

### 2.1.2 Relative Areas of the Anode and Cathode

The rate of corrosion is greater when the area of the cathode is larger. When the cathodic area is larger, the demand for electrons will be greater, and this results in increased rate of dissolution of metals at anodic regions. The corrosion of the anode is directly proportional with the ratio of the cathodic area to anodic area. Rapid and severe corrosion is observed if the anodic area is small due to heavy current density at the anodic area.

### 2.1.3 Purity of Metal

Presence of impurities leads to the formation of local electrochemical cells. In other words, the impurities present in a metal create heterogeneity, and thus galvanic cells are set up with distinct anodic and cathodic areas in the metal. The higher the percentage of impurity present in a metal, the faster is the rate of corrosion of the anodic metal. For instance, impurities such as lead and iron in zinc result in the formation of tiny electrochemical cells at the exposed part of the impurity, and the corrosion of zinc around the impurity takes place due to local action. Corrosion resistance of a metal may be improved by increasing its purity.

### 2.1.4 Physical State of the Metal

Metal components subjected to unevenly distributed stresses are easily corroded. Even in a pure metal, the areas under stress tend to be anodic and suffer corrosion. As an example, caustic embrittlement corrosion in a metal takes place in stressed parts such as bends, joints, and rivets in boilers.

### 2.1.5 Passivity or Passivation

The phenomenon in which a metal or an alloy exhibits much higher corrosion resistance than expected from its position in the electrochemical series is known as passivity or passivation. Formation of a very thin protective and invisible film around 0.0004 mm thick on the surface of the metal or an alloy makes it noble. One example is steel containing Ni and Cr. Chromium (Cr) forms a protective layer of $Cr_2O_3$ on steel, making it passive in oxidizing environments. Gold (Au) and platinum (Pt) are chemically very inert and hence show superior corrosion resistance properties. The elements or alloys can be formatted in a series with decreasing tendency of anode formation or nobility, as shown below:

Na > Mg & Mg alloys > Zn > Al > Cd > Fe > steel and cast iron > Pb > Sn > Cu > Ni > Cr > stainless steel > Ag > Ti > Au > Pt

### 2.1.6 Nature of the Corrosion Product

If the corrosion product is soluble or volatile in the corroding medium, then the underlying metal surface will be exposed readily, and corrosion occurs at a faster rate; however, if the corrosion product is insoluble in the corroding medium, forming a film at the surface, then the protective film formed tends to suppress further corrosion. If the corrosion product is oxide, the rate of corrosion mostly depends on the specific volume ratio; the greater the specific volume ratio, the lesser is the oxide corrosion rate.

### 2.1.7 Nature of the Oxide Film

Metals such as Mg, Ca, and Ba form oxides with volumes less than the volume of the metal. Hence, the oxide film formed is porous, through which oxygen can diffuse and bring about further corrosion. On the other hand, metals like Al, Cr, and Ni form oxides with volumes greater than that of metal, and the non-porous oxide film so formed protects the metal from further corrosion.

## 2.2 Nature of the Corroding Environment

### 2.2.1 Effect of Temperature

The rate of corrosion increases with increasing temperature, since the rate of chemical and electrochemical reactions and the rate of ions increase, which is why stress corrosion and intergranular corrosion are usually observed at high temperatures. Additionally, a passive metal may become active at a higher temperature.

### 2.2.2 Dissolved Oxygen Concentration and Formation of Oxygen Concentration Cells

The rate of corrosion increases with increasing supply of oxygen, which is the reason why the corrosivity of water decreases with temperature, since dissolved oxygen content decreases with temperature.

The regions where oxygen concentration is lesser become anodic and suffer corrosion. Corrosion often takes place under metal washers, where oxygen cannot diffuse readily. Similarly, buried pipelines and cables passing from one type of soil to another suffer corrosion due to differential aeration such as lead pipeline passing through clay and then through sand. The part of the lead pipeline that passes through clay gets corroded, since clay is less aerated than sand.

### 2.2.3 Nature of the Electrolyte

The nature of the electrolyte also influences the rate of corrosion. If the conductance of the electrolyte is high, for instance, the corrosion current is easily conducted, and hence the rate of corrosion is increased. Also, if the electrolyte consists of silicate ions, they form insoluble silicates and prevent further corrosion.

### 2.2.4 Presence of Corrosive Ions

If aggressive anions such as chlorides or sulfates are present in the medium, corrosion is accelerated, since they destroy the protective film, exposing the surface, leading to further corrosion taking place.

### 2.2.5 Flow Rate

High flow rates and liquid turbulence increase the corrosion, since they remove film formed by the corrosion products, exposing bare metal surfaces to corrosive chemicals.

### 2.2.6 Humidity

The rate and extent of corrosion increases with increasing humidity, which is why atmospheric corrosion of iron increases rapidly in the presence of moisture, while it is slow in dry air. This is due to the fact that moisture acts as the solvent for the oxygen in the air to furnish the electrolyte that is essential for setting up a corrosion cell. Thus, rusting of iron substantially increases when the relative humidity of air increases from 60% to 80%.

### 2.2.7 Effect of pH

The corrosion probability with respect to pH of the solution and the electrode potential of the metal can be determined with the help of Pourbaix diagrams, which clearly identify the zones of immunity, passivity, and the corrosion based on the pH and potential values. Thus, the corrosion rate of iron, for instance, can be reduced by increasing the pH of the solution by adding an alkali without disturbing the potential. Corrosion, particularly electrochemical corrosion, is largely depends on the pH of the medium. In general, acidic mediums are more corrosive than alkaline or neutral mediums.

Additionally, the iron would be immune from corrosion if the potential is changed to about –0.8V obtained from the Pourbaix diagram, which can be achieved by applying an external current. On the other hand, the corrosion rate of iron can also be reduced by moving it into the passivity region by applying a positive potential.

### 2.2.8 Presence of Impurities in the Atmosphere

Different types of suspended particles are present in the atmosphere. Some of them absorb moisture, leading to formation of galvanic cells, increasing the corrosion rate rapidly. Such particles are called active particles, e.g., NaCl, $(NH_4)_2SO_4$, sulfates, nitrates, etc. Presence of ammonia ($NH_3$) increases the corrosion rate of copper as follows:

$$Cu \longrightarrow Cu^{2+} + 2e^- \quad \text{(Eq. 36)}$$

$$Cu^{2+} + 4NH_3 \longrightarrow [Cu(NH_3)_4]^{2+} \quad \text{(Eq. 37)}$$

As the concentration of $Cu^{2+}$ ions decreases due to complexation, more copper metal dissolves to form $Cu^{2+}$ ions. In a similar way, dissolution of zinc (Zn) metal occurs faster in presence of ammonia ($NH_3$) as follows:

$$Zn \longrightarrow Zn^{2+} + 2e \quad \text{(Eq. 38)}$$

$$Zn^{2+} + 4NH_3 \longrightarrow [Zn(NH_3)_4]^{2+} \quad \text{(Eq. 39)}$$

Charcoals, aerosols, etc. increase the corrosion rate, indirectly acting as catalysts, which is why they are called inactive particles.

Another factor causing corrosion of metals and alloys is acid rain, which is usually caused by combustion of fuels containing sulfur. Gases such as $SO_x$ and $NO_x$ released into atmosphere due to these combustions combine with water vapor and humidity, forming $H_2SO_4$ and $HNO_3$, which fall down to earth with rains, corroding, for instance, historical artifacts that are metallic in character or made of bronze. While gases such as $SO_2$ act as corrosion accelerators, oxidizing gases such as ozone ($O_3$) show corrosion-preventive characteristics. Ozone reduces the activity of $SO_2$ and forms protective oxide layers on metal surfaces.

# 3

# Corrosion Mechanisms

Corrosion monitoring, identification of the causes, and application of preventive measures, as well as coming up with appropriate designs, all require first understanding of the electrochemical mechanisms of corrosion, followed by understanding the thermodynamic approach determining the corrosion tendency, polarization in a sense referring to corrosion process coming into equilibrium and analysis of corrosion rate, and finally passivity.

## 3.1 Direct Chemical Attack or Chemical or Dry Corrosion

Dry or chemical corrosion is corrosion due to the oxidation of metals due to chemical gases in the environment absent water vapor and humidity. Corrosion due to water vapor and humidity is considered aqueous corrosion and is included in the electrochemical corrosion category. Whenever corrosion takes place by direct chemical attack by gases like oxygen, nitrogen, and halogens, a solid film of the corrosion product is formed on the surface of the metal, which protects the metal from further corrosion. If a soluble or volatile corrosion product is formed, then the metal is exposed to further attack. For example, chlorine and iodine attack silver, generating a protective film of silver halide on the surface. On the other hand, stannic chloride formed on tin is volatile, and so corrosion is not prevented. Oxidation corrosion is brought about by direct action of oxygen at low or high temperatures on metals in the absence of moisture. Alkali metals such as Li, Na, K, etc. and alkaline earth metals such as Mg, Ca, Sn, etc. are readily oxidized at low temperatures. At high temperatures, almost all metals except Ag, Au, and Pt are oxidized. Alkali and alkaline earth metals on oxidation produce oxide deposits of smaller volume. This results in the formation of a porous layer through which oxygen can diffuse to bring about further attack of the metal. On the other hand, aluminum, tungsten, and molybdenum form oxide layers of greater volume than the metal from which they were produced. These non-porous, continuous, and coherent oxide films prevent the diffusion of oxygen, and hence the rate of further attack decreases with an increase in the thickness of the oxide film. Three different types of chemical or dry corrosion are usually observed:

i. Oxidation corrosion
ii. Corrosion by other gases
iii. Liquid metal corrosion

### 3.1.1 Oxidation Corrosion

In this type of corrosion, metals are oxidized to their oxides, producing four different types:

*i. Stable oxides*

In the cases of aluminum (Al), copper (Cu), etc., oxides developed on the surface of the metal are stable and impervious. The oxide appears as tightly adhering film to protect the underneath metal. Oxides are formed according to the following reactions:

$$Al + O_2 \longrightarrow Al_2O_3 \qquad \text{(Eq. 40)}$$

$$Cu + O_2 \longrightarrow CuO \qquad \text{(Eq. 41)}$$

Both the alumina ($Al_2O_3$) and copper oxide (CuO) act as protective coatings.

*ii. Unstable oxides*

In some cases, metal oxides, formed on the metal surface, decompose back to the metals and oxygen. Silver (Ag), gold (Au), and platinum (Pt) oxides are highly unstable, and hence they do not undergo oxidation corrosion. Thus, silver, gold, and platinum are highly stable and not susceptible to oxidation corrosion.

*iii. Volatile oxides*

Some metal oxides are volatile, and hence oxidation continues to take place until total metal is converted to the corresponding metal oxide. One such example is oxidation of molybdenum (Mo).

$$2Mo + 3O_2 \longrightarrow 2MoO_3 \text{ (volatile)} \qquad \text{(Eq. 42)}$$

*iv. Porous oxides*

The protective or non-protective nature of the oxide film is determined by a rule known as the Pilling-Bedworth rule. The ratio of the volume of the oxide formed to the volume of the metal consumed is called the Pilling-Bedworth ratio. Accordingly, if the specific volume of the oxide layer is greater than the volume of the metal, the oxide layer is protective and non-porous, e.g., $Al/Al_2O_3$, Cu/CuO, etc. So these metals are the least susceptible to oxidation corrosion. On the other hand, if the specific volume of the oxide formed on the surface is less than that of metal, the oxide film produced on the metal surface becomes porous, allowing continuous aeration through pores, and thereby helps continue the corrosion until all of the metal is exhausted. In other words, oxide layers are sufficiently stressed or strained, leading to the formation of cracks and pores, such as in the case of Li, Na, K, etc.

$$M + O_2 \longrightarrow M_2O \text{ or } MO \text{ or } M_2O_y \qquad \text{(Eq. 43)}$$

These metals are highly susceptible to corrosion. $M^{y+}$ is small in size and has a tendency to diffuse towards the surface at fast rate. $O^{2-}$ is large in size, and hence inward diffusion of $O^{2-}$ through oxide layer is slow. There are several possibilities of diffusion of ions:

i. metal ions may migrate outwards
ii. oxide ions may migrate inwards
iii. molecular oxygen may penetrate to the metal/ oxide interface
iv. both processes I and II occur simultaneously.

Thus, the smaller the specific volume, the greater is the rate of oxidation corrosion. Specific volume of the pore is defined as the ratio of the volumes of metal oxide to the volume of the metal:

*specific volume of pore = volume of metal oxide/volume of metal* (1)

Specific volume of Tungsten (W), chromium (Cr), and Nickel (Ni) are 3.6, 2.0, and 1.6, respectively. The rate of corrosion is least in the case of tungsten (W), and thus tungsten is stable even at high temperatures.

### 3.1.2 Corrosion by Other Gases

Corrosion also occurs with gases such as chlorine ($Cl_2$), hydrogen sulfide ($H_2S$), etc. For instance, silver (Ag) undergoes corrosion in presence of $Cl_2$ according to the following reaction:

$$2Ag + Cl_2 \longrightarrow 2AgCl \qquad \text{(Eq. 44)}$$

However, AgCl film is protective, and prevents further attack of chlorine on silver. Another example is the reaction of tin (Sn) with chlorine, forming $SnCl_4$, which is volatile and hence accelerates the corrosion of tin metal.

$$Sn + 2Cl_2 \longrightarrow SnCl_4 \qquad \text{(Eq. 45)}$$

### 3.1.3 Liquid Metal Corrosion

When a liquid metal is flowing over a solid metal surface through different temperature zones, there is a possibility that solid metal gets dissolved in the liquid metal at the high temperature zone, then gets deposited back again at low temperature zone. For example, sodium metal leads to corrosion of cadmium in nuclear reactors.

## 3.2 Electrochemical or Aqueous or Wet Corrosion

When corrosion occurs through the formation of galvanic cells, the phenomenon is called electrochemical corrosion. To complete the cell circuit, flow of ions is essential, and hence a medium called electrolyte is needed, which is mostly moisture or water. According to the electrochemical theory, the corrosion of a metal in aqueous solution is a two-step process, one involving oxidation and then reduction. It is known that two metals having different electrode potentials form a galvanic cell when they are immersed in a conducting solution. The electromotor force (emf) of the cell is given by the difference between the electrode potentials of the metals. When the electrodes are joined by a wire, electrons flow from the anode to the cathode. The oxidation reaction occurs at the anode, where the metal atoms lose their electrons to the environment and pass into the solution in the form of positive ions. Thus, there is a tendency at the anode to destroy the metal by dissolving it as ions. Hence, corrosion always occurs at anodic areas. The electrons released at the anode are conducted to the cathode and are responsible for various cathodic reactions such as electroplating or deposition of metals, hydrogen evolution, and oxygen absorption:

*i. Electroplating*

Metal ions at the cathode collect the electrons and deposit on the cathode surface:

$$Cu^{2+} + 2e^- \longrightarrow Cu \qquad \text{(Eq. 46)}$$

*ii. Liberation of Hydrogen*

In an acid solution, in the absence of oxygen, hydrogen ions accept electrons and hydrogen gas is formed:

$$2H^+ + 2e^- \longrightarrow H_2 \qquad \text{(Eq. 47)}$$

In a neutral or alkaline medium, in the absence of oxygen, hydrogen gas is liberated with the formation of $OH^-$ ions:

$$2H_2O + 2e^- \longrightarrow H_2 + 2OH^- \qquad \text{(Eq. 48)}$$

*iii. Oxygen Absorption*

In the presence of dissolved oxygen and in an acid medium, oxygen absorption reaction takes place:

$$4H^+ + O_2 + 4e^- \longrightarrow 2H_2O \qquad \text{(Eq. 49)}$$

In the presence of dissolved oxygen and in a neutral or weakly alkaline medium, $OH^-$ ions are formed.

$$2H_2O + O_2 + 4e^- \longrightarrow 4OH^- \qquad \text{(Eq. 50)}$$

The following reactions exemplify some of the anodic and cathodic reactions that take place during the corrosion of zinc and iron metals in acidic and basic conditions.

In acidic medium, zinc corrodes as follows:

$$Zn \longrightarrow Zn^{+2} + 2e^{-} \quad \text{(Eq. 51)}$$

With no oxygen, cathodic reaction is:

$$2H^{+} + 2e^{-} \longrightarrow H_2 \quad \text{(Eq. 52)}$$

With oxygen it is:

$$O_2 + 4H^{+} + 4e^{-} \longrightarrow 2H_2O \quad \text{(Eq. 53)}$$

In aerated neutral and basic conditions, iron corrodes as follows:

$$Fe \longrightarrow Fe^{+2} + 2e^{-} \quad \text{(Eq. 54)}$$

$$O_2 + 2H_2O + 4e^{-} \longrightarrow 4OH^{-} \quad \text{(Eq. 55)}$$

and the net reaction is:

$$2Fe + 2H_2O + O_2 \longrightarrow 2Fe^{+2} + 4OH^{-} \quad \text{(Eq. 56)}$$

leading to further reactions,

$$2Fe^{+2} + 4OH^{-} \longrightarrow 2Fe(OH)_2 \quad \text{(Eq. 57)}$$

$$2Fe(OH)_2 + H_2O + \tfrac{1}{2}O_2 \longrightarrow 2Fe(OH)_3 \quad \text{(Eq. 58)}$$

While it is easier to passivate iron at pH values between 10 and 12, especially via inhibitors adsorbed on iron's surface, it is difficult to do so below pH 8.

Thus, the essential requirements of electrochemical corrosion are formation of anodic and cathodic areas, electrical contact between the cathodic and anodic parts to enable the conduction of electrons, and an electrolyte through which the ions can diffuse or migrate, which is usually moisture. When these requirements are met, numerous galvanic cells are set up in presence of a conducting medium like water. Oxidation of metals occurs at the anode and the metal ions flow towards the cathode. $OH^{-}$ and $O^{-2}$ ions are formed at the cathode, these ions move toward the anode and the product, and as a result, metal oxide is formed somewhere between the cathode and anode.

A common general example to electrochemical corrosion is concentration cell corrosion, also called differential aeration corrosion, which manifests itself in different types of corrosion such as pitting, crevice, filling, underground soil corrosion, etc. In this type of corrosion, anodic and cathodic areas may be generated even in a perfectly homogeneous and pure metal due to different amounts of oxygen reaching different parts of the metal and forming oxygen concentration cells. In such circumstances, those areas that are exposed to greater amounts of air become cathodic, while the areas that are little exposed or not exposed to air become anodic and suffer corrosion.

Hence, an area covered with dirt, which is less accessible to air, becomes anodic and suffers corrosion. Since the anodic area is small and the cathodic area is large, corrosion is more concentrated at the anode, leading to the formation of a small hole on the surface of the metal, which is an intense local corrosion called pitting.

In another example, in a wire fence, areas where the wires cross are less accessible to air than the rest of the fence, and hence corrosion takes place at the wire crossings, which are anodic. In a similar way, iron corrodes under drops of water or salt solution. Areas covered by droplets, having less access to oxygen, become anodic with respect to the other areas that are freely exposed to air.

## 3.3 Differences between Chemical and Electrochemical Corrosion

The following are the general differences between dry (chemical) and wet (electrochemical) corrosion:

1. Chemical corrosion occurs in the dry state; electrochemical corrosion occurs in wet conditions in the presence of moisture or electrolyte.

2. Chemical corrosion involves the direct chemical attack by the environment; electrochemical corrosion involves the setting up of a huge number of tiny galvanic cells.
3. Chemical corrosion follows adsorption mechanism; electrochemical corrosion follows the mechanism of electrochemical reactions.
4. In chemical corrosion, even a homogenous metal surface will corrode; while in electrochemical corrosion, only heterogeneous metal surfaces or homogenous metal surfaces with bimetallic contact will corrode.
5. In chemical corrosion, corrosion products accumulate in the same spot where corrosion occurs; while in electrochemical corrosion, corrosion occurs at the anode and products gather at the cathode.
6. In chemical corrosion, uniform corrosion takes place; while in electrochemical corrosion, pitting corrosion is more frequent, especially when the anode area is small.
7. Chemical corrosion is a slow and a uniform process; electrochemical corrosion is a fast and non-uniform process.

# 4

# Corrosion Types

In the following chapter, the most common types of corrosion will be reviewed within two major categories, uniform and non-uniform or localized corrosion.

## 4.1 Uniform Corrosion

Uniform corrosion is the most common form of corrosion and causes the most metal loss, often leaving behind a scale or deposit; however, it is not deemed very dangerous since it occurs uniformly over the entire exposed surface of metal, leading to a certain predictable amount of metal thinning. For example, if a zinc plate is immersed in dilute sulfuric acid, the metal dissolves at the entire surface dipped in sulfuric acid.

Theoretical calculations performed to measure corrosion rate are based on the assumption that occurring corrosion is uniform corrosion, and thus metals exposed to other types of corrosion reveal corrosion long before the predicted time by the theoretical calculations. Uniform corrosion rate is calculated either as the unit weight loss per area per time in mg/dm²/day or the thickness loss of material per unit of time, which is commonly expressed as corrosion penetration rate (CPR) calculated via the following formula:

$$CPR = kw/\rho at \qquad (2)$$

where w is the weight loss after exposure time t

ρ and a represent the density and exposed specimen area

k is a constant and its magnitude depends on the system of units used.

CPR is usually expressed in two ways, either in millimeters per year (mm/yr) or in mils per year (mils/year or mpy). When expressed in mm/year, the terms given in the formula (2) above take the following values and units, K = 87.6 and w, ρ, a, and t are specified in units of mg, $g/cm^3$, $cm^2$, and hours, respectively; when expressed in mils/year or mpy, K = 534 and units are same as in the case of mm/year, except a is expressed in $inch^2$.

Inasmuch as there is an electric current associated with electrochemical corrosion reactions, corrosion rate can also be expressed in terms of corrosion current, or, more specifically, current density, that is, the current per unit surface area of material corroding, which is designated "i." Thus, the corrosion rate (r) is determined using the following expression in units of $mol/m^2.s$:

$$r = i/nF \qquad (3)$$

where n is the number of electrons associated with the ionization of each metal atom and F is equal to 96.500 C/mol.

Although electrochemical corrosion involves numerous microscopic galvanic cells, and thus cannot be entirely homogeneous, encompassing corrosive agents, such as the atmosphere, fresh water bodies such as lakes or rivers, or saltwater bodies such as the sea or soil, may induce corrosion on metallic structures that are homogenously exposed to these surrounding corrosive agents, which can be idealized as uniform corrosion, e.g., corrosion of iron sheets, tarnishing of silver plates, etc. Thus, atmospheric corrosion, corrosion due to fresh and salt waters, and underground corrosion will be reviewed in the uniform corrosion category as types of electrochemical corrosion, while high temperature corrosion will be reviewed later in the uniform corrosion category as a type of dry corrosion.

### 4.1.1 Atmospheric Corrosion

Atmospheric conditions can vary widely locally. Even geographical directions are important. For instance, a compact structure's east and south sides are less susceptible to corrosion than west and north, because east and south sides are dried faster with sun than west and north sides. However, the damaging effect of UV or sun radiations to the paints must also be considered, especially after getting wet, since the paints on the east and south sides will be affected more by the sunlight as well.

Salt content of the air increases with increasing altitude, resulting in more corrosion, while at low altitudes, forests and mountains slow down the wind speed, lessening the salt water content. Winds from the sea carry chloride ions that are usually very effective a few kilometers inland. In marine environments, the amount of chloride accumulated on metal surfaces from air is between 5 to 500 $mg/m^2/day$, while it can exceed 1500 $mg/m^2/day$ at the coast, and less than 5 $mg/m^2/day$ in rural inland areas.

Major air pollutant and corrosive sulfur dioxide gas in air originates from combustion reactions of petroleum and coal, which both contain sulfur and are very effective within a 3 km diameter. In urban areas, $SO_2$ accumulated on metal surfaces from air is between 10–80 $mg/m^2/day$, and in industrial areas, sometimes over 200 $mg/m^2/day$, while in rural areas, it is only 10 $mg/m^2/day$.

Solid precipitates originated from air that accumulated on the metal surface such as dust are hygroscopic and constitute an acidic environment, thus increasing corrosion.

For atmospheric corrosion to occur, an aqueous film of a certain thickness must be present on the metal surface. For the aqueous film to reach such thickness, relative humidity of the atmosphere must reach a critical humidity value. For steel surfaces in indoors, this critical humidity value is accepted generally as 60%, while outdoors it is 80%. It is calculated that there is 0.01 $g/m^2$ water present on the metal surface at critical relative humidity values, which increases up to 1 $g/m^2$ water at 100%relative humidity and up to 100 $g/m^2$ water when covered with rain water. On the other hand, if the aqueous film on the surface is too thick, corrosion is impeded, since diffusion of oxygen becomes more difficult. Thus, the thickness of the aqueous layer on the metal surface that causes the most corrosion is accepted around 150 μm. Thus, a surface that is exposed to wet and dry cycles or, in other words, that gets wet and dry alternately, is more susceptible to corrosion than a surface that is always wet, since solubility of oxygen in water is very low and progress of atmospheric corrosion depends on dissolved oxygen.

The aqueous layer may also contain sulfur dioxide, carbon dioxide, and chlorides, which all accelerate corrosion. Corrosion deposits on the metal surface and high relative humidity values such as 80% results in all sulfur dioxide ($SO_2$) to bind the metal surface. Sulfur dioxide first forms sulfur trioxide, anhydride of sulfuric acid, which produces sulfuric acid in the presence of water, leading to a decrease in pH below 4.

Temperature increase usually increases corrosion; however, it also dries the aqueous layer on the metal surface, and thus there is a temperature when the corrosion is the highest, and reduces thereafter. Below 0°C, corrosion can usually be omitted.

Atmospheric corrosion rate is the highest when the metal is first exposed to air, and then it decreases in time. For example, 60% of mild steel's corrosion takes place during the first year in a service life of 16 years. Thus, it is better if the metal is initially cured at an environment that is less corrosive than the atmosphere, and then it can be placed into the more corrosive environment, since the protective passive film that forms on the metal surface in the meantime can protect the metal from corrosion when exposed to air.

Steel's atmospheric corrosion rate varies from 5–10 μm/year in rural areas, up to 10–30 μm/year in marine environments, and up to 10–60 μm/year in industrial areas; while for zinc, it is 0.5–1 μm/year in rural, 0.5–2 μm/year in marine, and 1–10 μm/year in industrial areas; for aluminum, it is less than 0.1 μm/year in rural, 0.4–0.6 μm/year in marine, and 1 μm/year in industrial areas; and for copper, it is less than 1 μm/year in rural, 1–2 μm/year in marine, and 1–3 μm/year in industrial areas, respectively. Thus, steel alloys with copper and nickel as the alloying elements are resistant to corrosion for long periods unless chlorides are abundant in the environment. Therefore, electricity poles made of alloys of steel with copper and nickel can be carried in open air to far distances without being corroded. Such alloys are also less brittle than carbon steel and thus are more suitable to be carried such distances.

### 4.1.2 Corrosion in Water

*i. Corrosion in Seawater*

In the seawater, salt concentration varies from 32 g/L up to 36 g/L in tropical waters and away from the coasts. Of this 36 g, roughly around 20 g is Cl, 11 g is Na, 3 g is $SO_4$, 1 g is Mg, 0.5 g is Ca, 0.5 g is K, and the rest is bicarbonates, bromides, and strontium. Salt concentration (S) of sea is calculated by multiplying a constant with chloride percentage by weight:

$$\textit{Salt Concentration (g/kg)} = 1.80655 \times \textit{wt.of chlorides per liter of seawater(g/kg)} \quad (4)$$

Dissolved oxygen concentration decreases with salt concentration, and it is 11.0 mg/L at 0°C for 36g/L chloride concentration, compared to 14.6 mg/L in fresh water. There are differences in terms of aeration and thus in terms of dissolved oxygen concentration reaching different parts of a steel structure embedded in the sea. The part of the steel structure over the seawater is exposed to atmospheric corrosion, which is like a marine environment with high chloride concentrations. The part at the sea level is always wetted with the waves and is also called the "splash zone," and the corrosion deposits are constantly washed away and cannot form a protective layer on the metal surface. The corrosion rate is the highest at this level. Right below the sea level, structure gets wet and dry, a protective layer can form promptly, and thus the corrosion rate is relatively low. Inside the sea in the regions near the sea surface, corrosion rate is also high, since the structure is always wet and dissolved oxygen concentration is high. At the lowest level of the structure in the deeper sea level, the dissolved oxygen concentration is the lowest, and thus the corrosion rate is the lowest as well.

Seawater's pH is 8 on average; however, it is higher near the surface of the sea since the plants that live at the surface of the sea get sunlight and use dissolved carbon dioxide for photosynthesis, thus reducing the pH. At deep waters, due to formation of carbon dioxide and hydrogen sulfide originating from rotting dead bodies of living organisms, pH is around 7.5.

High conductivity of seawater causes formation of macro corrosion cells in addition to the microcells. Low ohmic resistance of the electrolyte results in the large ratio of the cathodic area to anodic area leading to severe pitting corrosion. Secondly, high chloride concentrations prevent reformation of passive films on the surfaces of metals that can normally passivate, e.g., aluminum and iron, making them susceptible to corrosion in the seawater.

Calcium and magnesium ions present in the seawater can precipitate on the metal surface, which is accelerated with the formation of hydroxide ions at the cathode, and especially when cathodic protection is applied. Therefore, while the required current for cathodic protection is high during the first several days, less is needed after the formation of the protective layer, which consists of 57% calcium carbonate, 19% iron oxides and hydroxides, 8% silicates, and 16% magnesium hydroxide, calcium sulfate, and others.

In general, temperature increases corrosion rate in seawater as well; however, since it also leads to quick formation of the protective layer and decreases the dissolved oxygen concentration, a peak of maximum corrosion is reached at around 80°C, and the corrosion rate reduces thereafter.

Some plants and animals that live in the sea adhere to the protective coating on the metal surface, leading to "fouling effect," which reduces the diffusion of oxygen to the metal surface and at the same time forms an acidic environment due to their rotten dead bodies, leading to increased corrosion as a result. These living organisms damage the paint coatings as well, unless the paint is poisonous.

Mobility of the metallic structures in the sea also affects the corrosion rate; for example, ships in seas with strong waves are more susceptible to corrosion, as it is clear in the case of application of cathodic protection, since the cathodic protection current need doubles when the ship is in motion, compared to being anchored at a port.

### *ii. Corrosion in Fresh Waters*

Dissolved oxygen concentration is higher in fresh waters compared to the seawater since it decreases with increasing salt concentration, and thus it is very dangerous to use pipes made of steel and copper in cold water or cooling water systems, for instance. However, dissolved oxygen concentration is relatively very low in closed circulated hot water systems due to the decreasing dissolved oxygen concentration with increasing temperature; thus it is not appropriate to replace radiator waters, since it will cause the oxygen to be replenished, and so oxygen scavengers such as sulfites or hydrazine are used to totally remove the oxygen. The amount of dissolved oxygen in distilled water at 0°C is 14.6 mg/L, which decreases down to

2.8 mg/L 80°C. Therefore, although the corrosion rate doubles with every 10°C increase in temperature, due to the reduction in dissolved oxygen concentration, a peak is reached at around 80°C, as it is the case in seawater, and the corrosion rate reduces thereafter until the temperature reaches 100°C, at which points water evaporates, resulting in a dissolved oxygen amount of 0, and aqueous corrosion ceases.

Substantial corrosion takes place for pH values lower than 4 even in the absence of oxygen, since the primary cathodic reaction at that pH level will be hydrogen reduction.

Corrosion is impeded due to water hardness precipitating on metal surfaces in the form of calcium carbonate along with the other corrosion products present at the metal surface at pH values higher than the saturation pH ($pH_s$):

$$Ca(HCO_3)_2 \longrightarrow CaCO_3 + H_2O + CO_2 \qquad \text{(Eq. 59)}$$

If pH is lower than $pH_s$, then the existing calcium carbonate precipitates may also dissolve back into the solution in the form of bicarbonates. The difference between both pH values is called the Langelier index, which is used to determine whether a protective shell is formed or not:

$$\textit{Langelier index (L)} = pH - pHs, \qquad (5)$$

while for values of $L > 0$, $CaCO_3$ precipitation occurs; for values of $L < 0$, it does not.

### 4.1.3 Underground or Soil Corrosion

Underground corrosion occurs for many reasons, such as galvanic effects, damaged coatings, different degrees of aeration and oxygen concentration, microorganisms present in the soil, differences in the nature and content of the soil, moisture content or humidity of the soil, electrolytes present in the soil and terrain's resistivity, redox potential of the terrain, acidity and pH of the soil, stray currents, and interference effects.

For aqueous, electrochemical, or wet corrosion to occur, a sufficient amount of dissolved oxygen is required, in addition to the electrolytic medium, which is regularly 8 ppm or 5.6 ml/L in water. Although it is easier for dissolved oxygen and water to reach the embedded metal structures in soils with large grains and gravels, it is difficult in soils with clay. Regions with less oxygen become the anode, and with more oxygen, become the cathode. As a result, in the case of buried pipelines and cables passing from one type of soil to another, the part of the pipeline that passes through the soil with clay, for instance, acts as the anode, while the part that passes through the soil with sand or sand with large grains and gravels, for instance, acts as the cathode due to differential aeration. When part of a pipeline is replaced with a new one, the new one becomes anode and the old one becomes the cathode, initiating corrosion due to potential difference between the two. Additionally, clay particles that adhere to pipelines cause potential difference and cause corrosion as well. Also, oxygen diffusion is greater in filled lands containing numerous air pockets than in natural ones, bringing about severe corrosion due to differential aeration. Another example is when there are different structures above the pipeline on the ground, e.g., asphalt vs. open ground, which results in the pipeline under the less permeable asphalt to become anode due to having less access to oxygen, and the surroundings become the cathode. Further, in wells, the region of the pipe embedded in water becomes the anode, since there is less oxygen in water and the part of the pipe right above the water that is in contact with air becomes the cathode.

In the presence of excess pesticides, germicides, or a large amount of organic matter, metals form soluble complexes, resulting in sufficient concentration difference of metal ions, resulting in several concentration cells, leading to severe soil corrosion. In water-logged areas, anaerobic bacteria becomes active, causing severe deterioration of soil in that area.

Soils that have low resistivity have high corrosivity. Low resistivity of the soil is due to the presence of moisture and dissolved electrolytes. These two factors promote corrosion.

Humidity is required for underground corrosion to occur, forming the electrolytic medium. In soils with mostly sand and gravel, water drains well, while in soils with clay, soil remains wet for long periods. Ground resistivity decreases with increasing soil humidity up to 20%–30% of the relative humidity, and remains constant afterwards. Ground resistivity decreases with temperature as well, while it increases substantially below 0°C. Ground resistivity also depends on the soil type, e.g., percentages of clay and silt, and on the dissolved ions present in the soil. Consequently, if ground resistivity is less than 1000 ohm. cm, it is considered very conductive and corrosive; if between 1000 and 3000 ohm.cm, it is accepted as corrosive, if between 3000 and 10000 ohm.cm, it is medium corrosive, and if above 10000 ohm.cm, it is considered only slightly corrosive. Werner's four electrode method is a common way to measure the resistivity of soils at the surface of the earth.

Regularly pH of soil is between 5 and 8, and does not have any effect on the corrosion leaving effect of the oxygen concentration as the primary determinant of corrosion. However, if pH of the soil is less than 5, which happens usually due to rotting organic materials or acid rains, numerous concentration cells cause corrosion, and protective coating of corrosion deposits or $CaCO_3$ cannot form on the metal surface.

Redox potential also gives an idea about corrosion. Terrain potentials lower than 100 mV indicate severely corrosive environments, while potential values between 100–200 mV indicate corrosive environments, values between 200–400 mV indicate mildly corrosive environments, and values higher than 400 mV indicate only slightly corrosive environments. Redox potential is usually measured using platinum electrodes and a pH-meter. A platinum electrode is placed in the environment and the potential difference with that of another reference electrode is measured and inserted into the following formula:

$$E_{redox} = E_{Pt} + E_{ref} + 60\,(pH-7) \tag{6}$$

$E_{Pt}$ is the potential of platinum electrode, and $E_{ref}$ is the potential of reference electrode with respect to standard hydrogen electrode (SHE), e.g., for saturated $Cu/CuSO_4$ electrode (CSE), it is 316 mV compared to SHE.

Steel structures placed underground are usually expected to have a service life of 50 to 100 years; however, if environmental factors result in corrosive factors, this may lead the service life to be less than 50 years, and preventive measures have to be taken. If the anode and cathode are near one another and if the pH of the soil is higher than 5, the corrosion products are settled at the metal surface in the form of rust, leading to a reduction in the corrosion rate in time. However, especially in the case of pipelines buried underground, anode and cathode may be very far away from one another, and since the oxidized metal ions move towards the cathode, and hydroxide ions form at the cathode move towards the anode, rust forms somewhere in between, away from both the anode and the cathode, and thus the layer of corrosion products cannot protect the metal surface.

It is usually not economical to change the properties of the soil to prevent corrosion; however, sometimes pipelines are embedded in clean sands covered with ditches or they are coated with protective coatings, but most commonly, cathodic protection is implemented.

### 4.1.4 High Temperature Corrosion

High temperature corrosion is a type of dry corrosion. Oxidation of metals and alloys at high temperatures is one of the fundamental problems in energy production technology, as it is in gas turbines. As a result of oxidations at high temperatures, eutectic mixtures of low melting points can also form between the different oxides of the metals as they occur in the case of hot vapor turbines.

Oxides generally are semiconductors. ZnO, for instance, is less protective when $Zn^{+2}$ ions are replaced with $Al^{+3}$, since $Al^{+3}$ will donate the extra electron it has to the ZnO film, making it an n-type semiconductor, while it will be more protective when $Zn^{+2}$ ions are replaced with $Li^{+}$. Spinel oxides have less conductivity, such as dark red $FeO.Al_2O_3$, which has 18% Al and is used instead of chromium steels at low temperatures, because at high temperatures, less noble element in the alloy oxidizes at the surface, leaving the nobler metal behind, making the alloy rich in the nobler element. Thus, if nickel is alloyed with gold, for instance, gold forms a protective coating at the surface at high temperatures.

In petroleum refining, lighter products separate during the refining process, which leads to an increase in the concentration of sulfur and vanadium, creating problems that surface during energy production. Vanadium compounds that have low melting points lead most alloys to be able to be used only for a few months unless Co-Cr alloys are used as structural materials. Another prevention method is increasing the melting points of vanadium compounds with the addition of calcium and magnesium compounds. On the other hand, extracting vanadium from within the system is an expensive method. Sulfates and pyrosulfates also lead to the formation of low melting point compounds; however, calcium and magnesium compounds bind 90% of the sulfur.

Melted salts increase conductivity, and presence of oxidizing and reducing agents prevent polarization, which both accelerate corrosion at high temperature environments.

Nonmetals such as concrete easily break during freezing and melting, especially if there are salt and acidic anhydrides such as carbon dioxide and sulfur dioxide present in the environment. Acidic anhydrides become acids with rain waters and become even more concentrated due to the presence of bacteria, which are the primary cause of wooden materials to break apart. Polymers and plastics can also break due to UV light, ozone, solvents, and vapors. Use of carbon black pigment can prevent the UV damage, and it is used for that purpose in automobile tires.

## 4.2 Non-Uniform Corrosion

Non-uniform corrosion or localized corrosive attack is a type of electrochemical corrosion. There are many types of non-uniform corrosion that occur, primarily depending on the type of the metal and the characteristics of the corrosive environment.

### 4.2.1 Galvanic Corrosion

Galvanic corrosion is a common type of corrosion that occurs when two metals or alloys with different compositions are electrically coupled while exposed to an electrolyte, e.g., a regular atmosphere that leads to the formation of aqueous layers on metal surfaces. Electrical coupling occurs when two metals are immersed in an electrically conducting solution and are in contact through an electrical connection, resulting in the flow of electrons due to formation of a potential difference. The less noble or more reactive metal in the particular environment that has a more negative electrode potential becomes the anode, and thus goes into solution or corrodes, as in the case of steel pipes that are connected to copper plumbing. Similarly, steel components in the vicinity of a junction of copper and steel tubing also corrode in a domestic water heater. Another example of galvanic corrosion is when an aqueous solution of a more noble metal flows over an active metal. For instance, if waters flow over a copper metal, then over steel, copper ions that are dissolved will cause corrosion of iron, even if they are in very small concentrations. Other examples of galvanic corrosion are aluminum-copper and mild steel-stainless steel connections in marine environments.

In corrosion cells, metals are never in an equilibrium state with their ions; thus, it is more suitable to use galvanic series that list electrode potentials in seawater, rather than standard electrode potentials indicating the relative reactivities of a number of metals and alloys. When two alloys are coupled in seawater, the one lower in the series will experience corrosion. Some of the alloys in the series are grouped in brackets. Generally, the base metal is the same for these bracketed alloys, and there is little danger of corrosion if alloys within a single bracket are coupled. Additionally, some alloys are listed twice in the series, such as nickel and the stainless steels in both their active and passive states.

Corrosion rate in a galvanic cell depends on the difference between the potentials of the anode and the cathode, which reduces by time due to polarization. Other factors affecting galvanic corrosion rate are the conductivity of the electrolyte and the area ratio between the cathode and the anode. A more conductive electrolyte causes galvanic corrosion to occur in a larger area, resulting in less damage, while an electrolyte that has low conductivity leads to severe corrosion, where two metals connect to one another. Also, if cathodic to anodic area ratio is high, anodic current density increases substantially, leading to severe corrosion in a small area. For this reason, corrosion rate depends on current density, that is, the current per unit area of corroding surface, and not simply the current. Thus, a high current density results for the anode when its area is small relative to that of the cathode. Such an example is when a copper riveted steel plaque and a steel riveted copper plaque are placed in seawater. Steel riveted copper plaques corrode very fast due to the large area of copper cathode, indicating that it would be more effective to paint over the cathodic areas when such metals are in contact. A number of measures may be taken to significantly reduce the effects of galvanic corrosion:

1. Avoiding galvanic couples, especially if these metals are further apart in the galvanic series.
2. If coupling of dissimilar metals is necessary, metals that are close together in the galvanic series should be chosen and the cathodic/anodic area ratio must below.
3. During the design, easier replacement of the anodic material must be predicted and planned, or the thickness of the anodic material must be increased.
4. Connections between metals must be insulated with insulated flanges, and metal surfaces must be insulated with paints or coatings.
5. A third, more anodic metal can be electrically connected to the other two, leading to a form of cathodic protection.
6. If the system is a closed system, inhibitors should be used.

### 4.2.2 Crevice Corrosion

Crevice corrosion occurs when metals are in contact with non-metallic substances like wood, plastic, rubber, etc., resulting in presence of a crevice in between. Diffusion of the electrolytes to the crevice is difficult and slow, which leads to a concentration differential inside the crevice with that of outside in terms of oxygen concentration. Inside the crevice, oxygen concentration is low, since it is consumed by corrosion reactions, while it is abundant on the outside. Thus, the metal in contact with oxygen-rich solution at the outside acts as cathode, and the metal that is in contact with the solution within the crevice having little or no oxygen acts as anode. Since the metal inside the crevice corrodes as the anode, areas around the crevice do not corrode.

The crevice, such as one between different metallic objects, e.g., bolts, nuts, and rivets, that is in contact with liquids must be wide enough for the solution to penetrate, yet narrow enough for stagnancy; usually, the width is several thousandths of an inch. A common example is a scarcely aerated hatched portion of the riveted joint becoming anode being susceptible to corrosion, whereas the free part becomes cathode and is protected.

After oxygen has been depleted within the crevice, oxidation of the metal occurs. Electrons produced by the oxidation of the metal are conducted through the metal to adjacent external regions, where they are consumed by the reduction reactions, primarily reduction of oxygen. In time, the electrolyte inside the crevice becomes rich in $Fe^{+2}$ ions, since it is a stagnant solution, and this attracts the $Cl^-$ ions from the surroundings to inside the crevice to create an electrochemical balance. Iron chloride ($FeCl_2$) easily hydrolyzes to iron hydroxide [$Fe(OH)_2$] precipitate and hydrochloric acid (HCl). Consequently, crevice corrosion becomes an autocatalytic reaction, leading to a critical corrosion state, with up to 10 times more chloride concentration and pH values falling down to 2–3. Many metals and alloys that can passivate well or the ones that can precipitate in the form of hydroxides, such as nuts made of 18–8 stainless steels, are more sensitive to crevice corrosion because their protective films are often destroyed by the $H^+$ and $Cl^-$ ions.

$$Fe \longrightarrow Fe^{+2} + 2e^{-} \quad \text{(Eq. 60)}$$

$$Fe^{+2} + 2Cl^{-} \longrightarrow FeCl_2 \quad \text{(Eq. 61)}$$

$$FeCl_2 + 2H_2O \longrightarrow Fe(OH)_2 + 2H^{+} + 2Cl^{-} \quad \text{(Eq. 62)}$$

Chloride ions have an accelerating effect in crevice corrosion; thus, in environments with no chlorides present, crevice corrosion may occur after a longer time, such as a year. Crevice corrosion is also promoted by changes in the pH of the water, as well as presence of other aggressive anions similar to $Cl^{-}$ in the stagnant solution in the crevices. Following measures can be taken to prevent crevice corrosion:

1. Welding should be preferred instead of nuts or rivets, e.g., using welded instead of riveted or bolted joints.
2. Nonabsorbing gaskets should be used when possible.
3. Areas where metal plaques meet must be insulated with welding.
4. During the design, containers or containment vessels that may contain liquids must be designed to avoid stagnant areas and ensure complete drainage, and thus no corners must remain that cannot be cleaned and washed.
5. Such containers must also be checked regularly to ensure there are no precipitates or accumulations of deposits.
6. Materials that can stay wet, such as wood or plastics, must not be in contact with the metals.

### 4.2.3 Pitting Corrosion

In pitting corrosion, a small area of the metal surface is affected and formation of cavities takes place, while the remainder of the surface remains unaffected. The formation of cavities or pits usually occurs due to a localized surface defect, e.g., breakdown or cracking of the protective film due to a scratch, for instance. The surface diameter of the pits is more or less same as that of their depth. They ordinarily penetrate from the top of a horizontal surface downward in a nearly vertical direction. It is supposed that gravity causes the pits to grow downward and the solution at the tip becomes more concentrated and dense as the pit growth progresses. Therefore, pitting corrosion is concentrated on narrow gaps in static solutions caused primarily by $Cl^-$ and $Br^-$ ions in neutral conditions. It is replaced by uniform corrosion in lower pH values.

Metal loss is very small in pitting corrosion, but materials can be punctured due to cavities or pits growing very fast, leading to holes in the metal, resulting in leakage of fluids that causes mechanical failures; thus, it is an extremely insidious and dangerous type of corrosion. It often goes undetected due to the small size of the pits, which are also commonly covered with corrosion products.

All metals or alloys that have passivation properties are as sensitive to pitting corrosion as they are to crevice corrosion, such as aluminum alloys and stainless steels. Even mild steel is more resistant to pitting corrosion than stainless steels; however, alloying stainless steels with about 2% molybdenum enhances their resistance significantly.

The extent of pitting corrosion cannot be predicted by weight loss measurements; thus, it is measured by the number of pits and the average pit depth, or the pitting factor. Pitting factor is calculated as the maximum pit depth, which is the average depth of the five biggest pits that are calculated statistically based on probability calculations divided by the average thickness loss that is calculated from the weight loss measurements that take into account only uniform corrosion.

$$\textit{pitting factor} = \frac{\begin{array}{c}\textit{maximum pit depth}\\ \textit{(average depth of five biggest pits)}\end{array}}{\textit{average thickness loss}} \qquad (7)$$

The mechanism for pitting is probably the same as for crevice corrosion, in that oxidation of metal occurs in a narrow area within the pit itself, constituting the anode, with complementary reduction at a large area around the pit at the surface, constituting the cathode. Produced metal ions attract chloride ions from the surroundings, increasing the hydrogen ion concentrations inside the pit and lowering the pH, while oxygen gas is reduced at the surroundings of the pit, constituting the cathodic reaction. In time, the pit's mouth is covered with corrosion products, preventing chloride ions to move into the pit, resulting in a decrease in the corrosion rate. Thus, pitting corrosion can only occur in stagnant solutions, such as in pipelines and storage tanks at locations where flow rate or fluid motion is slow.

An example of pitting corrosion happens in the case of a water droplet resting on a metal surface. The metal surface that is covered by the droplet acts as the anode due to less access to oxygen, and suffers corrosion. The metal surface that is uncovered acts as the cathode, due to high oxygen concentration. As the anodic area is small compared to the cathodic area, more and more metal is removed at the same spot. Thus, a small hole is formed on the surface of the metal, leading to pitting corrosion.

Presence of impurities on the surface of a metal also leads to pitting corrosion. In fact, it has been observed that specimens having polished surfaces display a greater resistance to pitting corrosion. In such cases, the metal underneath the impurity, like scales, act as an anode, and the surroundings become the cathode due to different oxygen concentrations.

Using sufficient amount of inhibitors and cathodic protection are useful methods to prevent pitting corrosion; however, insufficient inhibitor dosage would lead to a higher ratio of cathodic surface area to anodic surface area, resulting in more cathodic currents concentrating in a small anodic surface area, leading to more severe corrosion.

### 4.2.4 Selective Leaching or Selective Corrosion

Selective leaching occurs in solid solution alloys when one element or constituent is preferentially removed as a consequence of corrosion processes, resulting in the loss of luster and surface texture. The most common example is the dezincification of brass, in which zinc is selectively leached from the brass alloy that has 70% copper and 30% zinc. The mechanical properties of the alloy are significantly impaired, since only a porous mass of copper remains in the region that has been dezincified. Additionally, the material changes color from yellow to copper red. Brass's resistance to selective corrosion decreases with increasing zinc percentage in the alloy. The best composition would be when the zinc percentage is less than 15%, along with 1% of tin anad trace amounts of arsenic, antimony, or phosphorus as inhibitors. Further, stagnant solution conditions are more suitable for selective corrosion to occur. Selective corrosion does not require the presence of oxygen; copper and zinc corrode first, and while zinc ions stay in solution, copper ions reduce cathodically and deposit back onto the metal, leading to the formation of a porous structure. Another example of selective leaching is graphitization, which occurs in gray cast iron, where 2% to 4% carbon in the alloy becomes cathode and iron becomes the anode, resulting in iron leaching away, leaving graphite carbon behind in the structure. The same does not occur in white cast iron, since carbon is not free in the structure as it is in gray cast iron. Selective leaching may also occur in other alloy systems in which aluminum, iron, cobalt, chromium, and other elements are vulnerable to preferential removal.

### 4.2.5 Filiform Corrosion

Filiform corrosion occurs in the metals such as aluminum, steel, zinc, etc. that are commonly coated with paint or rubber. Filiform corrosion usually progresses in the form of an irregular shaped line. Its initiation point is blue-green in color, whereas the filaments are brown. Filiform corrosion is a surface phenomenon, and does not affect the strength of the metal. The most important factor causing filiform corrosion is the relative humidity of the atmosphere. At or above 90% relative humidity, filiform corrosion rate increases substantially.

Filiform corrosion can be considered a type of crevice corrosion, since it occurs underneath the paint or another coating at the metal surface. Corrosion initiates at a weak point of the coating, where oxygen and water can enter. At this point, oxygen concentration is the maximum, while it becomes less and less concentrated along the path corrosion proceeds. As a result of the corrosion, metal hydroxides and hydrogen ions form, and thus the terminal point of corrosion underneath the coating, the furthest location from the weak point of the coating, where oxygen does enter, becomes the best environment for corrosion to proceed even further, due to low oxygen concentration and low pH. Filiform corrosion is prevented if the coating or paint is waterproof and strong.

### 4.2.6 Erosion Corrosion

The term erosion corrosion is a combined action of mechanical abrasion and wear on the surface of metal as a consequence of fluid motion and corrosion. Erosion corrosion can usually be identified by surface grooves, troughs, and waves with contours that are characteristic of the flow of the fluid. As the main reason for this type of corrosion is due to turbulent flow of the liquid, it is also called as turbulence corrosion. Also, a solution is more erosive when bubbles and suspended particulate solids are present.

At first, a corrosion product is formed, e.g., the protective oxide layer of the metal, which erodes away especially due to abrasive action caused by the turbulent fluid movement of corrosive fluids at high velocities along with particle impingement, leaving an exposed bare metal surface. In other words, first, the oxide film or film of corrosion products breaks or deteriorates mechanically due to abrasion or due to flow of liquid or gas, followed by the chemical or electrochemical corrosion process that begins under conditions of corrosive medium, leading to formation of pits in the direction of flow of the liquid or gas at the metal surface. The turbulence effect of the flowing liquid that yields erosion corrosion is usually due to a pit that was previously formed on the metal surface due to regular corrosion processes.

The nature of metallic surface, nature of the fluid, flow rate, and turbulent flow conditions are the major factors. Although faster flow rate translates to more economy, erosion corrosion also increases with increasing flow rates. Thus, the speed is usually not increased beyond 1.2 m/s in steel pipes and 1.5 m/s in copper pipes, and stays between 7.5 m/s to 9.0 m/s in stainless steel pipes. When the flow rates are increases beyond these levels, such as up to 2.5 m/s to 3.0 m/s in regular steel pipes, the turbulence effect increases, and countercurrents form. One solution here can be using a wider pipe.

Another trade-off is at the pipe thickness. Erosion corrosion occurs in heat exchangers even in the case of liquids that are not corrosive, since thicker pipes are preferred, which translates to 25% less inner surface area for the pipe, leading to an increase in the flow rate. Thus, in such cases, usually pipes made of aluminum and nickel alloys of copper, e.g., 90–10 copper-nickel alloy pipes, are used instead of steel pipes, which allow the flow rate to be increased up to 3 m/s carrying seawater. Using iron as an alloying element in trace amounts also helps the formation of iron oxide film that is resistant toward seawater.

Most metals and their alloys are susceptible to erosion corrosion, especially alloys that passivate by forming a protective surface film. Additionally, relatively soft metals such as copper and lead are more susceptible to erosion corrosion in comparison to stainless steel, aluminum, and other metals that passivate. If the coating composed of corrosion products is not capable of continuous and rapid reformation, erosion corrosion may be severe. Erosion corrosion mainly occurs in equipment with fast flowing liquids such as in the case of pipelines, especially at bends and elbows, at locations where there are abrupt changes in pipe diameter, positions where the fluid changes direction or flow suddenly becomes turbulent, and also in turbine blades, pumps, propellers, valves, centrifuges, mixers, heat exchangers, condensers, ducts, turbine equipment, etc. Measures that can be taken to prevent erosion corrosion are:

1. Choosing a structural material that is resistant to wears and erosion corrosion.

2. Changing the design of the component to eliminate fluid turbulence and impingement effects.
3. Providing barrier wear resistant coatings.
4. Using wider pipes, reducing the flow rate.
5. Strengthening regions, e.g., valves, that are susceptible to erosion corrosion by increasing metal thickness.
6. Removing the particulates and bubbles by precipitating solid particles that are present in the solution.
7. Reducing temperature.
8. Cathodic protection.

### 4.2.7 Cavitation Corrosion

Cavitation corrosion is a type of erosion corrosion commonly occurs in hydraulic turbines, ship propellers, etc. When the flowing liquid contains gas or vapor, this pressurized gas explodes at obstacles on the metal surface when in contact and results in damage. The mechanism is such that while flowing very fast, at some locations, pressure becomes low, creating a vacuum effect, causing water to evaporate forming vapor or allowing the dissolved gases in the liquid to separate, which explode at rough locations at the metal surface, leading to formation of pits. In other words, these tiny bubbles deflate at locations where flow rate is reduced, such as nearby rough areas at the metal surface, creating a vacuum effect at the metal surface forming pits.

There is also corrosion with cavitations, which is different than the cavitation corrosion. Corrosion with cavitations occurs very commonly, and can be prevented by administering inhibitors or cathodic protection, while cavitation corrosion can only be prevented during the designing stage.

### 4.2.8 Abrasion Corrosion

Abrasion corrosion occurs in metals that vibrate under a load, and also in metals that are in contact that move relatively to one another, even if the friction motion of vibration is as small as $10^{-10}$ cm. It is common to have abrasion corrosion together with fatigue as in the case of metal implantations placed in the human body for medical reasons. Relative motion of two surfaces in opposite directions results in the removal of the protective metal oxide film, exposing the bare metal surface to corrosive agents. When galvanized materials' surfaces are worn away, they corrode fast; thus, they must be stored separately and ventilated well. They should be also lubricated and fixed not to move. Further, relative motion of two surfaces in abrasion corrosion leads to small metal pieces separating from the metal due to mechanical friction, and these pieces get oxidized easily. Aluminum alloys are usually very sensitive to abrasion corrosion; thus, they are first lubricated and then loaded to ships. Compressors, automobiles, railway transportation, etc. are commonly protected against abrasion corrosion via lubrication as well.

### 4.2.9 Stress Corrosion

Stress corrosion, also commonly called stress corrosion cracking (SCC), results from the combined action of mechanical stress, such as static or applied tensile stress, and a corrosive environment. Stress may result from applied forces during manufacture, fabrication, heat treatment, etc., or locked-in residual stress. Metal components are subjected to unevenly distributed stresses during the manufacturing process. Further, various treatments of metals and alloys such as cold working or quenching, bending, and pressing introduce uneven stress. The electrode potential thus varies from one point to another. Therefore, corrosion takes place so as to minimize the stress. Areas under great stress act as the anode, while areas not under stress act as the cathode.

Some materials that are virtually inert in a particular corrosive medium become susceptible to this form of corrosion when a stress is applied, since while normally corrosion products can form a protective coating on the metal surface to prevent further corrosion, they cannot while under stress. Small cracks form and then propagate in a direction perpendicular to the stress, eventually leading to a mechanical failure. Failure behavior is characteristic of that for a brittle material, even though the metal alloy is intrinsically ductile. Furthermore, cracks may form at relatively low stress levels, significantly below the tensile strength. The stress that produces stress corrosion cracking need not be externally applied; it may be a residual one that results from rapid temperature changes and uneven contraction, or for two-phase alloys in which each phase has a different coefficient of expansion. Also, gaseous and solid corrosion products that are entrapped internally can give rise to internal stresses. This type of corrosion leads to either intergranular or transgranular cracks in the metal. The intergranular cracking proceeds along grain boundaries, whereas transgranular cracking proceeds along individual grains. The metallic surface remains virtually unattacked in this type of corrosion, while fine cracks that have extensive branching gradually increase inside. Stress corrosion cracks are usually more pronounced in special corrosive conditions in which corrosion products are dissolved and the preventive layer is not reformed.

Pitting corrosion and intergranular corrosion increase the stress, causing stress corrosion cracking as a result. Corrosion of head and point portions of a nail indicates that they have been acting as anode to the middle portion. Actually, the head and the point portions were put under stress during the manufacture. In the case of iron-wire hammered at the middle, corrosion takes place at the hammered part, and results in breaking of the wire into two pieces.

Most alloys are susceptible to stress corrosion in specific environments, especially at moderate stress levels. For instance, carbon steel is susceptible in strong alkaline solutions and when nitrates are present in the environment, most stainless steels are susceptible in solutions containing chloride ions, and brass such as brass equipment used for agriculture are susceptible in nitrate solutions and in ammonia. Regular steels get sensitive to stress corrosion cracking as their carbon content gets lower than 0.1%. Also, some structures in which high strength cast alloys are used are susceptible to stress corrosion unless the environment is away from the sea. Specially designed steels are sensitive to stress corrosion cracking due to dissolved oxygen above 300°C. Regular steels are exposed to stress corrosion cracking due to hydroxides such as NaOH and KOH and $H_2S$ over 100°C, while high strength low alloys of steel are susceptible to stress corrosion cracking due to $H_2S$ over 20°C.

At high temperatures, even very low concentrations of chlorides can cause stress corrosion cracking in austenitic steels, while they are more resistant if their nickel content is above 10%. Additions of even small amounts of molybdenum, nitrogen, and silicon increase their resistance against stress corrosion cracking. Austenitic steels get sensitive to stress corrosion cracking due to precipitation of chromium carbide at grain boundaries, or in other words, intergranular corrosion. The same is observed for steels in nitrates and over 75°C.

Ferritic steels that have 18% to 20% chromium content are resistant to stress corrosion cracking, while their nickel and copper content lessen their resistance. In the absence of nickel, ferritic steels are very resistant to stress corrosion cracking.

Even small amounts of water at room temperatures may cause hydrogen embrittlement in martensitic steels leading to stress corrosion cracking, which is more pronounced (likely to occur) in the case of specially designed martensitic steels.

Aluminum alloys that have many alloying elements such as copper, magnesium, silicon, and zinc are susceptible to stress corrosion cracking via intergranular corrosion at corrosive environments coupled with high stress. In general, high strength aluminum alloys are susceptible to stress corrosion cracking due to chlorides and at temperatures above 20°C. Specifically, high strength aluminum alloys that are contact with seawater are susceptible to stress corrosion cracking. Since ships are wanted to be constructed light, high strength aluminum-silicon alloys are used; however, these alloys are not resistant to stress corrosion, especially in icy and cold waters. Therefore, medium strength alloys usually are more suitable.

Titanium and zirconium alloys are susceptible to stress corrosion cracking due to melted chloride salts at temperatures above the melting point of such salts. Titanium alloys are susceptible to stress corrosion cracking due to liquid $N_2O_4$ at temperatures over 50°C.

Copper alloys are susceptible to stress corrosion cracking due to ammonia and humidity at temperatures over 20°C, and their resistance even lessens with addition of alloying elements such as arsenic, phosphorous, antimony, and silicon. Phosphated pure copper is usually considered resistant to stress corrosion cracking, but not immune to intergranular corrosion. Both copper and zinc are susceptible to corrosion in ammonia solution, since ammonia dissolves out Cu and Zn as $[Cu(NH_3)_4]^{2+}$ and $[Zn(NH_3)_4]^{2+}$, respectively, creating tensile stress on the metal surface.

Pure magnesium is resistant to stress corrosion cracking, while its alloys containing more than 1.5% aluminum, especially if exposed to temperatures between 50°C and 200°C for long periods, are sensitive mostly due to precipitation of $Al_3Mg_2$ at grain boundaries.

Pure nickel is resistant to stress corrosion cracking even in chloride solutions, as it is against halogens that are not oxidizers, while acidic chlorides of iron, copper, and mercury cause severe corrosion. Certain nickel alloys may be susceptible to stress corrosion cracking due to intergranular corrosion in high temperature aqueous solutions.

One of the ways to prevent stress corrosion cracking is to reduce the magnitude of the stress via heat treatments to anneal out any residual thermal stresses and/or via reducing loads, which can be done either by reducing the external load or increasing the cross-sectional area perpendicular to the applied stress. Materials can be made more resistant to the tensile stress via annealing such as keeping brass at 300°C for 1 hour or annealing stainless steel at 500°C. For low carbon steels, this operation is done between 595°C and 650°C, while for austenitic steels it is done between 815°C and 930°C. Another prevention technique is to employ phosphates, such as inorganic or organic inhibitors, to prevent corrosion in amounts just needed, because exceeding amounts may lead to pitting and other types of corrosion. Additionally, coating the metal is also an effective method of prevention. When none of these can be done, alloys susceptible to stress corrosion cracking are replaced with the resistant ones, e.g., 304 type stainless steel with nickel-rich inconel alloy. Also, carbon steels, low carbon steels, and decarburized steels, which are less expensive, are more resistant to stress corrosion cracking than stainless steels, which is contrary to their resistance towards uniform corrosion, in which stainless steels are more resistant. Thus, heat transfer units or heat exchangers that are in contact with seawater are usually made of carbon steel to prevent stress corrosion cracking. Cathodic protection is also an effective method; however, if stress corrosion cracking is due to hydrogen embrittlement, then applied cathodic current would only increase corrosion.

### 4.2.10 Intergranular Corrosion

Metals solidify in the forms of grains consisting of crystalline units, e.g., iron has a cubic centered unit crystalline structure, while austenitic steels have face centered unit cubic structure. Boundaries of these grains meet the surface at different places and become active under certain conditions, resulting in localized corrosion attacks. Additionally, crystalline structure is irregular between grains at grain boundaries, and thus is more susceptible to corrosion. Intergranular corrosion occurs preferentially along these grain boundaries. The net result is that a macroscopic specimen disintegrates along its grain boundaries, leading to a reduction in the mechanical strength of the metal. In some cases, the metal converts to powder, due to disintegration into separate grains.

Impurities usually accumulate at grain boundaries, such as a little amount of iron in aluminum. Iron dissolves very little in aluminum and thus accumulates at the grain boundaries, constituting an irregularity. As accumulation of impurities at grain boundaries causes intergranular corrosion, sometimes their absence also does, e.g., absence of chromium, which is the major alloying element in steel, at grain boundaries results in corrosion, since chromium protects steel from corrosion when its percentage is 12% or above in the alloy. Another example is 18–8 steel, which regularly has 0.2% carbon that can be reduced down to 0.08% with easy procedures; however, specific methods are required for more purification. When 18–8 steel containing Cr and Ni is heated to between 500 and 800°C and especially at 650°C, chromium reacts with carbon, which is the other alloying element in steel when in amounts more than 0.02%, forming $C_{23}C_6$, which is not soluble in steel and thus accumulates at the grain boundaries, resulting in lower chromium concentration at grain boundaries compared to the bulk. Thus, regions where $Cr_{23}C_6$ precipitate become anodic and are vulnerable to corrosion. This process in stainless steel is known as sensitization and the phenomenon is known as intergranular corrosion. The same phenomenon occurs in 304 stainless steels as well, since they contain 0.06% to 0.08% of carbon. Although chromium in the bulk of the alloy moves to the grain boundaries where it is less in concentration in the solid solution, this motion is very slow and cannot prevent the corrosion. Intergranular corrosion is the reason why stainless steels cannot be welded unless the welded steel material is not too thick and the welding duration is short, since then material cools rapidly, not allowing chromium carbide sufficient time to form. For the same reason, if stainless steels are to be welded, electricity welding would be more appropriate. This type of failure is commonly known as weld-decay.

Intergranular corrosion is also observed in many non-ferrous metals, e.g., in precipitation hardened duralumin consisting of Al and Cu. Stainless steels may be protected from intergranular corrosion by the following measures:

1. Subjecting the sensitized material to a high-temperature heat treatment in which all the chromium carbide particles are re-dissolved. For this

reasons, steel is heated up to 1100°C then cooled rapidly in water or in appropriate oil. At such high temperatures, chromium carbide is in the solid solution, and thus it can be homogenously distributed in the alloy.

2. Alloying the stainless steel with other metals such as titanium, niobium, or columbium that have greater tendencies to form carbides than does chromium so that the Cr remains in solid solution, as in the case of 321 and 347 steel alloys. The carbides of these alloying elements, however, melt at higher temperatures than chromium carbide, and thus may accumulate at grain boundaries in the form of a long line at both sides of welding area, while chromium carbides remain in solution phase.
3. Lowering the carbon content below 0.03 wt. C% so that carbide formation is minimal as in the case of ELC (extra low carbon) steels such as 304L steel.

### 4.2.11 Caustic Embrittlement

Caustic embrittlement is the phenomenon during which the boiler material becomes brittle due to the accumulation of caustic substances. It is a very dangerous form of stress corrosion, occurring at high temperatures in mild steel boiler metals exposed to alkaline solutions and resulting in the failure of the metal. Boiler water usually contains a small proportion of $Na_2CO_3$. In high pressure boilers, this breaks up to give NaOH and makes the boiler water more alkaline:

$$Na_2CO_3 + H_2O \longrightarrow 2NaOH + CO_2 \qquad \text{(Eq. 63)}$$

This alkaline boiler water flows into the minute hair cracks and crevices such as rivet holes by capillary action. There, the water evaporates and the concentration of caustic soda increases progressively. The concentrated alkali dissolves the metallic iron as sodium ferrate in crevices, cracks, etc. where the metal is stressed. Consecutively, sodium ferrate decomposes to $Fe_3O_4$, giving rise to strong tensile stresses on the steel surface:

$$Fe + NaOH \longrightarrow Na_2FeO_2 \quad \text{(Eq. 64)}$$

$$3Na_2FeO_2 + 4H_2O \longrightarrow 6NaOH + Fe_3O_4 + H_2 \quad \text{(Eq. 65)}$$

$$6Na_2FeO_2 + 6H_2O + O_2 \longrightarrow 12NaOH + 2Fe_3O_4 \quad \text{(Eq. 66)}$$

The regenerated caustic alkali helps further dissolution of iron, leading to the brittlement of boiler parts, particularly stressed parts of the boiler such as bends, joints, and rivets, even causing total failure of the boiler. Caustic embrittlement can be prevented by employing inhibitors or by applying protective coatings on the surface. Additionally, boiler tank water can be buffered with phosphates and volatile ammine compounds, so that pH cannot increase substantially at cracks.

### 4.2.12 Hydrogen Embrittlement

Various metal alloys, especially some steels, experience a significant reduction in ductility and tensile strength when atomic hydrogen (H) penetrates into their crystalline structure. Often the reason of such penetration is corrosion reactions in general, overprotection phenomenon in cathodic protection, electroplating, pickling operations in general, and pickling of steels in sulfuric acid in specific, high-temperature operations such as heat treatments and welding with a wet electrode in presence of hydrogen-bearing atmospheres including water vapor or sour gas environments. All of these processes result in formation of hydrogen atoms at the metal surface, leading some of these hydrogen atoms to be adsorbed and diffuse into the holes in the metal and combine therein to form hydrogen gas and accumulate, causing an increase in the volume and pressure, since a hydrogen molecule is bigger than two hydrogen atoms, and hydrogen molecules cannot diffuse back out as hydrogen atoms can. This phenomenon is aptly referred to as hydrogen embrittlement; the terms hydrogen-induced cracking and hydrogen stress cracking are sometimes also used.

Strictly speaking, hydrogen embrittlement is a type of failure in response to applied or residual tensile stresses. Brittle fracture occurs catastrophically as cracks grow and rapidly propagate. Hydrogen in its atomic form (H), as opposed to the molecular form ($H_2$), diffuses interstitially through the crystal lattice, and concentrations as low as several parts per million of it can lead to cracking. If metal is not under stress, some of the hydrogen atoms that did not combine to form hydrogen molecules diffuse back out.

Hydrogen embrittlement is similar to stress corrosion in that a normally ductile metals experience brittle fracture when exposed to both tensile stress and a corrosive atmosphere. However, these two phenomena may be distinguished on the basis of their interactions with applied electric currents. Whereas cathodic protection reduces or causes a cessation of stress corrosion, it may, on the other hand, lead to the initiation or enhancement of hydrogen embrittlement. Furthermore, hydrogen-induced cracks are most often transgranular, although intergranular fracture is observed for some alloy systems. A number of mechanisms have been proposed to explain hydrogen embrittlement; most of them are based on the interference of dislocation motion by the dissolved hydrogen.

Presence of what are termed "poisons," such as sulfur containing $H_2S$ or arsenic compounds, accelerate hydrogen embrittlement. These substances retard the formation of molecular hydrogen and thereby increase the residence time of atomic hydrogen on the metal surface. Hydrogen sulfide, probably the most aggressive poison, is found in petroleum fluids, natural gas, oil-well brines, and geothermal fluids. National Association of Corrosion Engineers (NACE) Standards of Material Specifications (MR 0175) define sour gas environments that may lead to hydrogen embrittlement as liquids containing water and $H_2S$ that have partial pressures of more than 0.0035 bar. Atomic hydrogen resulting from an electrochemical reaction between the metal and the $H_2S$ containing medium enters the steel at the corroding surface. These hydrogen atoms mostly accumulate at grain boundaries, resulting in the formation of hydrogen gas over time, leading to an increase in the volume and pressure. If this occurs at sites close to the surface, it leads to hydrogen blistering, and if occurs at inner sites, it leads to staircase-like cracks independent of structural stress. However, if there is also high structural stress in the environment where hydrogen gas formation occurs, cracks form perpendicular to the direction of the structural stress. Energy released by the exothermic reaction of hydrogen atoms forming hydrogen gas causes stress, contributing to the hydrogen induced cracking process as well.

Low carbon and low alloy steels that are typically used in pipelines may be susceptible to cracking when exposed to corrosive $H_2S$ containing environments, the severity of which depends on the hydrogen concentration, structure of the steel alloy, stress density, temperature, and environmental conditions. The primary reason for the corrosion in petroleum pipelines is water, which absorbs $O_2$, $H_2S$, and $CO_2$. In the case of iron and steel, there are holes in interstitial cubic centered unit structure; thus they can accept a foreign ion or atom in its structure that hydrogen atoms can diffuse. During pickling operations or welding with a wet electrode, the following reaction produces hydrogen atoms diffusing especially into α-iron:

$$Fe + H_2O \longrightarrow FeO + 2H \qquad \text{(Eq. 67)}$$

High-strength steels are susceptible to hydrogen embrittlement, and increasing strength tends to enhance the material's susceptibility. Martensitic steels are especially vulnerable to this type of failure, while bainitic, ferritic, and spheroiditic steels are more resilient. Furthermore, face centered cubic (FCC) alloys such as austenitic stainless steels and alloys of copper, aluminum, and nickel are relatively resistant to hydrogen embrittlement, mainly because of their inherently high ductilities. However, strain hardening these alloys will enhance their susceptibility to embrittlement.

Reducing MnS inclusions as well as adding calcium and rare earth metals such as cerium to the alloy increases the resistance to hydrogen induced corrosion. Other measures that can be taken to prevent hydrogen embrittlement are:

1. Metal can be heated up to 100°C–150°C so that the absorbed hydrogen atoms diffuse back out of the interstitial crystalline structure; in other words, the alloy is "baked" at an elevated temperature to drive out any dissolved hydrogen.
2. Nickel and molybdenum can be added to the high strength steel alloys, or the alloy may be substituted with an alloy that is more resistant to hydrogen embrittlement.
3. Operations leading to hydrogen formation at the metal surface must be avoided, such as wet welding or overprotection, which is the application of a higher-than-needed cathodic protection potential.
4. Tensile strength of the alloy can be reduced via heat treatment.

### 4.2.13 Corrosion Fatigue

Fatigue is defined as a term for fracture of structures subjected to dynamic and fluctuating stresses, as in the case of bridges, aircrafts, and machine components. Metals that are under varying dynamic stresses of loading and unloading can become fatigued and crack with the effect of corrosion under small stresses that are considerably lower than the tensile or yield strength for a static load. The term fatigue is used because this type of failure normally occurs after a lengthy period of repeated stresses or strain cycling. Cyclic stresses may be axial (tension-compression), flexural (bending), or torsional (twisting) in nature. The nature of the stress cycles will influence the fatigue behavior; for example, lowering the load application frequency leads to longer periods during which the opened crack is in contact with the environment and to a reduction in the fatigue life.

Corrosion fatigue is dependent on several environmental factors such as temperature, pH, humidity, extent of aeration, etc. The physical properties of metals and their corrosion resistant nature also influence the corrosion fatigue. Many high quality steels have shorter fatigue life in the moist air as compared to dry air, and are less resistant to fatigue corrosion than carbon steels in general. Similarly, fatigue resistance of aluminum and bronzes decreases considerably in sea water. Because of fatigue, even in conditions where no corrosion takes place, steel's tensile strength can be halved. Only in salt water, tensile strength is 6% to 7% less than normal. The biggest reduction in the tensile strength occurs when effects of fresh water, salt water, and humid atmosphere are all present together. Fatigue corrosion is commonly observed in ship propellers. The load on the propellers constantly varies due to changing speed of the ship and since seawater is a corrosive environment, fatigue corrosion occurs. Another example is hot water pipes. Changes in the temperature of the water cause expansion and contraction of the pipes, leading to varying loads and stresses resulting in fatigue corrosion.

The main reason of corrosion fatigue is the development of cracks on the metallic surface due to scratches, corrosion pits, etc. These pits may have formed as a result of chemical reactions between the environment and material, which serve as points of stress concentration and therefore as crack nucleation sites. The bottom of such pits or cracks has been found to have more negative potential and hence works as the anode of a galvanic cell. The crack propagation then occurs due to stress, and is accelerated by corrosive process.

It is believed that in absence of corrosive medium, if a metal is stressed below its fatigue limit, it can undergo infinite number of cycles without fracture. Fatigue is important inasmuch as it is the single largest cause of failure in metals, estimated to comprise approximately 90% of all metallic failures; polymers and ceramics, with the exception of glasses, are also susceptible to this type of failure. Fatigue is catastrophic in the sense that it occurs very suddenly and without warning leading to brittlements even in normally ductile metals. The process occurs by the initiation and propagation of cracks, and ordinarily, the fracture surface is perpendicular to the direction of an applied tensile stress. Simple fracture is the separation of a body into two or more pieces in response to an imposed stress that is static, constant, or slowly changes with time in temperatures that are low relative to the melting point of the material. The applied stress may be tensile, compressive, shear, or torsional. Two fracture modes are possible: ductile and brittle. Ductile materials typically exhibit substantial plastic deformation with high energy absorption before fracture, while there is normally little or no plastic deformation with low energy absorption accompanying a brittle fracture. Ductile fracture is almost always preferred for two reasons: first, brittle fracture occurs suddenly and catastrophically without any warning as a result of rapid and spontaneous crack propagation. On the other hand, for ductile fracture, the presence of plastic deformation gives warning that the fracture is imminent, allowing preventive measures to be taken. Secondly, more strain energy is required to induce ductile fracture, inasmuch as ductile materials are generally tougher. Under the action of applied tensile stress, most metal alloys are ductile, whereas ceramics are notably brittle, and polymers may exhibit both types of fracture. For most brittle crystalline materials, the crack propagation corresponds to the successive and repeated breaking of atomic bonds along specific crystallographic planes, which is termed cleavage. On the other hand, in some alloys, crack propagation is along grain boundaries, and such fracture is termed intergranular.

The effect of fatigue corrosion can be reduced by coating the metal with zinc, chromium, nickel, or copper. Other measures that can be taken to prevent corrosion fatigue are:

1. Reducing the corrosive effects of the environment
2. Using a material resistant to corrosion
3. Using inhibitors
4. Cathodic protection
5. Taking measures during designing stage

### 4.2.14 Fretting Corrosion

Fretting corrosion is a physicochemical phenomenon that occurs at pressed contacts such as at a gear or a ball bearing on the rotating mile or axle in alternating loading conditions. Fretting corrosion occurs at the contact surface between pairs of closely contacting machine components that are not intended to move relative to each other, but do, however, move, due to component deflections where the relative motion ranges from a few micrometers up to 250 μm. Fretting fatigue caused by fretting corrosion induces a significant reduction of fatigue strength and consequently leads to unexpected failures even at very low stresses during service. Due to fretting corrosion, fatigue life of the part can reduce 3 to 6 times.

### 4.2.15 Stray-current and Interference Corrosion

In electric traction systems, such as in electric trains, electric current can leak into adjacent conducting structures. These stray currents go into the surrounding metallic structures and cause corrosion that is known as stray-current corrosion. This type of corrosion can lead to pitting and serious damages to underground structures. Railway systems that work with direct current, high voltage transmission lines that carry direct current and welding machines generate stray currents that escape into the earth or the terrains. For instance, a pipeline system that goes parallel to an underground subway system may be susceptible to corrosion. The direct current source's positive pole is connected to the subway train, while the negative pole is connected to the railway. Places where the stray currents enter the neighboring pipeline become the cathode, while the places where the stray currents exit the pipeline to go back to the direct current source become the anode and corrode.

Interference corrosion due to stray currents in the surrounding metallic structures occurs due to changes in potential fields both in the positive and negative directions that are created by cathodic protection systems. Two major types of interference corrosion are anodic and cathodic interference corrosion. Interference corrosion will be reviewed in detail in section 8.9, "Interference Effects of Cathodic Protection Systems."

### 4.2.16 Waterline Corrosion

Waterline corrosion occurs due to the difference of oxygen concentrations close to the water surface and below, e.g., an iron pole in stagnant seawater. The position near the waterline is well aerated and acts as the cathode. Areas deep inside are anodic since the oxygen concentration is less. Corrosion takes place in the anodic areas, and reduction of $O_2$ to $OH^-$ ions occurs at the cathodic regions. $Fe^{2+}$ ions produced at the anode and $OH^-$ ions produced at the cathode interact to yield $Fe(OH)_2$, which is further oxidized by dissolved oxygen to rust.

### 4.2.17 Microbial or Biocorrosion

The deterioration of materials through the involvement of microorganisms is known as microbial corrosion or biocorrosion or microbially influenced corrosion (MIC), which is due to microbial activities such as adsorbing of microorganisms on metal surfaces forming colonies, producing polymeric materials out of their cell structures, leading to a bio-film accelerating anodic and cathodic corrosion reactions. Microbial corrosion also accelerates corrosion activity of corrosion cells formed due to differences in aeration. Microbial corrosion is mostly encountered in cooling water systems, especially at locations where the flow is stagnant, while underground microbial corrosion takes place due to organic compounds present in the soil, when redox potential of the soil is low, indicating an anaerobic environment, and in temperatures less than 40°C and in pH values between 5 and 9.

To check whether microbial corrosion is taking place, a few drops of HCl can be added to the sample and if the distinct $H_2S$ smell is present, microbial corrosion can be said to have occurred. Petroleum, natural gas, waste water treatment, and transportation industries are susceptible to microbial corrosion. In specific, microbial corrosion of metal pipelines and equipment used for drilling, transportation, and storage in petroleum industry is mostly related to sulfate reducing bacteria (SRB). Sulfate reducing bacteria (SRB) reduce sulfate to produce energy producing toxic and corrosive $H_2S$ gas. Electron needed to reduce sulfate are provided from lactate, hydrogen, or other compounds. Along with sodium, chloride, magnesium, calcium and sulfate, petroleum reserves also contain hydrocarbon compounds and injection waters, which provide a suitable environment for sulfate reducing bacteria to develop. Due to activity of sulfate reducing bacteria, pitting corrosion occurs in metal equipment; injection wells are clogged with corrosion products such as iron sulfide, and produced biofilms lead to safety issues.

The difference between iron corrosion in the presence of bacteria and without is that without bacteria, surface iron dissolves to $Fe^{2+}$ and electrons stay at metal surface; water is reduced to protons and protons are reduced to hydrogen gas with the present electrons. In such conditions, corrosion is limited because cathode is polarized with the hydrogen gas present at the cathode. With sulfate reducing bacteria (SRB), however, hydrogen sulfate is taken away from the metal surface while being reduced, and this leads to an increase in the anodic dissolution of the metal and production of FeS and $Fe(OH)_2$. In systems with low $Fe^{2+}$ concentrations, temporary and adhesive iron sulfide film reduces the corrosion current density; however, this protective iron sulfide film under the sulfate reducing bacteria (SRB) biofilm does not form in systems with high $Fe^{2+}$ concentrations. As $Fe^{2+}$ concentrations increase, the number of sulfate reducing bacteria (SRB) also increases, since sulfate reducing bacteria (SRB) requires the presence of iron to reproduce, leading to an increase in corrosion current densities and shifts in corrosion potentials to anodic values.

Microorganisms play an important role in extracting minerals through bioleaching; for instance, thiobacillus ferrooxidant is a widely used organism for leaching sulfide minerals. Aerobic bacteria such as thiobacillusthiooxidant oxidize any sulfur compound and sulfur up to sulfates, forming sulfuric acid as a result.

$$2S + 3O_2 + 2H_2O \longrightarrow 2H_2SO_4 \qquad \text{(Eq. 68)}$$

On the contrary, anaerobic bacteria such as desulfovibrio that live at 25 to 30°C and at a pH range of 6 to 7.5, leading to a redox potential of about –100 mV, reduces the sulfates to sulfide ions, which accelerate iron's corrosion:

$$SO_4^{2-} + 8H^+ + 8e^- \longrightarrow S^{2-} + 4H_2O \qquad \text{(Eq. 69)}$$

$$4\,Fe \longrightarrow 4Fe^{2+} + 8e^- \qquad \text{(Eq. 70)}$$

$$4\,Fe + SO_4^{2-} + 8H^+ \longrightarrow 3Fe^{2+} + FeS + 4H_2O \qquad \text{(Eq. 71)}$$

Biocorrosion can be prevented effectively using inhibitors. Gluteraldehyde, for instance, prevents SRB-induced corrosion on stainless steels, metals, plastics, and glass materials. Gluteraldehyde interacts with the cell wall, cell membrane, and proteins in cytoplasm of the bacteria, preventing the exchange of materials in and outside of the cell, thus killing them. Usually, 50 ppm to 200 ppm concentrations of gluteraldehyde are sufficient; however, amounts of gluteraldehyde or formaldehyde used may need to be increased depending on the type of bacteria and conditions of the surroundings, which may cause problems, since environmental protection agencies allow only up to 50 ppm gluteraldehyde to be used. Another prevention method is to use inorganic chemicals, e.g., chlorine gas, chlorine dioxide, ozone, and bromine, or to use organic compounds, e.g., quaternary ammonium compounds or aldehydes that kill SRB. Among other measures that can be taken are pH adjustments, periodical chlorination, or use of organometallic tin compounds. However, bacteria can adapt to such measures in time; thus, removing the trace elements such as zinc and vanadium that the bacteria depend on to live appears to be the best solution.

# 5

# Thermodynamics of Corrosion

It is common to refer to thermodynamic parameters such as electromotor forces of common metals listed in standard emf and galvanic series when assessing a metal's susceptibility to corrosion. The standard emf and galvanic series include the list of reduction electrode potentials and are simply rankings of metallic materials on the basis of their tendency to corrode when coupled to other metals. For the standard emf series, ranking is based on the magnitude of the voltage generated when the standard cell of a metal is coupled to the standard hydrogen electrode at 25°C (77°F). The metals and alloys near the top of the series are cathodic, unreactive, noble, or chemically inert, e.g., gold and platinum, whereas those at the bottom are anodic, active, and more susceptible to oxidation and corrosion, e.g., sodium and potassium.

It should be noted, however, that although these potentials may be used to determine spontaneous reaction directions, they provide no information as to corrosion rates. That is, even though a $\Delta V$ potential computed for a specific corrosion situation is a relatively large positive number, the reaction may occur at only an insignificantly slow rate.

Additionally, emf half-cell potentials are thermodynamic parameters that relate to systems at equilibrium only. However, real corroding systems are not at equilibrium; there is always a flow of electrons from the anode to the cathode corresponding to the short-circuiting of the electrochemical cells. Furthermore, these half-cell potentials only represent the magnitude of a driving force, or the tendency for the occurrence of the particular half-cell reaction, and not the entire process. A more realistic and practical ranking is provided by the galvanic series, which represents the relative reactivities of a number of metals and commercial alloys, e.g., 316 stainless steel, 304 stainless steel, inconel, monel, bronzes, aluminum alloys, etc, in seawater with no voltages provided.

## 5.1 Gibbs Free Energy ($\Delta G$)

Even though emf series was generated under highly idealized conditions and has limited utility, it nevertheless indicates the relative reactivities of the metals and Gibbs free energy, and electromotor force formula is commonly used to assess the spontaneity of a chemical system:

$$\Delta G = -nF\varepsilon \tag{8}$$

where n is the number of grams or moles of electrons flowing through the corrosion cell, F (Faraday constant) is the charge of 1 gram or mole of electron and equals to 96.494 Coulomb, and E is the electromotor force of the corrosion cell, which can be calculated using the Nernst equation:

$$E = E^{\circ} - \frac{RT}{nF}\ln\left(\frac{activities\ of\ products}{activities\ of\ reactants}\right) \tag{9}$$

where R is gas constant that is 8.314 Joule/degree.mole, T is temperature in Kelvin, $E^{\circ}$ is the electromotor force in standard conditions, while activities are the effective concentrations of the reactants and products that is calculated via the following formula:

$$a = \gamma . c \qquad (10)$$

where c is the real concentration and $\gamma$ is the activity coefficient.

Activity coefficient of solids, electrons, and of species whose concentration does not change, e.g., solvents such as water, is equal to 1. For gases, partial pressure values are used in the place of activity coefficients, and the RT/nF is then converted to 0.0592 L.atm/mol.K at standard conditions.

For instance, in the case of galvanic corrosion, cell progresses via the following net reaction:

$$Zn + Cu^{2+} \longrightarrow Zn^{2+} + Cu \qquad \text{(Eq. 72)}$$

Activities of the solid zinc reactant ($a_{Zn}$) and the solid copper product ($a_{Cu}$) are equal to 1. Half-cell electrode potentials are calculated via Nernst equation:

$$Zn \longrightarrow Zn^{2+} + 2e^{-} \text{ (oxidation reaction)} \qquad \text{(Eq. 73)}$$

E° of zinc's oxidation reaction is 0.763V and number of electrons exchanged is 2. Thus,

$$E_{(Zn)} = 0.763 - 0.0592/2 \log a^{2+}_{Zn} \qquad (11)$$

$$Cu^{2+} + 2e^{-} \longrightarrow Cu \text{ (reduction reaction)} \qquad \text{(Eq. 74)}$$

E° of copper's reduction reaction is 0.363V and number of electrons exchanged is 2. Thus,

$$E_{(Cu^{2+})} = 0.363 - 0.0592/2 \log (1/a^{2+}_{Cu}) \qquad (12)$$

Consequently, the combined cell potential is as follows:

$$\varepsilon_{(Cu/Zn)} = 0.763 + 0.363 - 0.0592/2 \log (a^{2+}_{Zn} / a^{2+}_{Cu}) \qquad (13)$$

and if the activities of copper and zinc ions are the same, they would cancel each other out, resulting in the last term of the formula 13 to be 0 and $\varepsilon_{(Cu/Zn)}$ equaling to 1.126 V, which is a positive value, leading to a negative free energy value when inserted into formula 8, indicating that zinc dipped into copper sulfate solution has a tendency to corrode under standard conditions.

## 5.2 Passivity

Metals such as chromium, nickel, titanium, aluminum, magnesium, and iron that are above hydrogen in the electrochemical or galvanic series, and thus expected to corrode, do not, due to the oxide layers formed on their surfaces, resulting in passivity as observed in potential-pH diagrams. These metals automatically get passivated without application of any external current under appropriate conditions. Auto-passivation occurs when the corrosion current ($i_{corr}$) is higher than the current needed to passivate the metal. Hence, materials can be protected from corrosion by using alloys that have very low passivation currents or by adding oxidizing agents that are easily reducible, which are also called passivators, or, more commonly, inhibitors. Passivators or inhibitors are reduced electrochemically producing a corrosion current ($i_{corr}$) high enough, which surpasses the current needed to passivate the metal and preventing corrosion.

Examples of such alloys that allow passivation are chromium or chromium-nickel steels, nickel-chromium, copper-nickel, and titanium alloys, while such inhibitors are chromates, nitrites, molybdates, wolframates, ferrates, etc. In the case of using such alloys that have very low passivation currents, it is important to have appropriate compositions; for chromium-iron alloys, chromium amount should not be lower than 13%, to have passivation characteristics very similar to pure chromium; for copper-nickel alloys, alloy must be 50% to 60% nickel; and for silica-iron alloys, silica amount should not be lower than 14.5% for the best passivation properties.

If corrosion products form a protective film on the metal surface, corrosion rate is reduced. Effectiveness of this film depends on the solubility of the corrosion products, adhesive properties of these products on the metal surface, permeability, electrical resistance, and mechanical strength properties of the resulting coating. The best coatings form when the crystalline structure of the metal and the oxide film match. Since metals usually crystallize in cubic units, oxides, which also produce cubic crystalline structures, such as $Al_2O_3$ and $Fe_3O_4$, are better. Formation of cubic crystalline magnetite ($Fe_3O_4$), which has also low solubility and low electrical resistance, helps prevention of corrosion in boiler tanks.

Magnetite ($Fe_3O_4$) is the simplest of spinel oxides that have a formula of $RO.R'_2O_3$, where +2 charged constituent R can also be nickel or cobalt and +3 charged constituent R′ can be aluminum, chromium, or iron. In stainless steels, R is $Fe^{2+}$, while R′ is $Cr^{+3}$. In the case of 300 series chromium-nickel stainless steels, the protective oxide film consists of a spinel oxide, in which R is a mixed constituent of both $Ni^{2+}$ and $Fe^{2+}$ resulting in (NiO. FeO), while R′ is $Cr^{+3}$. In iron-aluminum alloy that has 18% aluminum, R is $Fe^{2+}$, while R′ is $Al^{3+}$. Although this alloy is resistant to corrosion, it is difficult to process, and it is also very weak at high temperatures. Single oxides may also form good protective coatings such as $Al_2O_3$ and BeO.

Since stainless steels are resistant to corrosion due to passivation, at environments where there is no oxygen, such as in some boiler tanks and reactors, use of cheaper regular steel alloys would be more appropriate, since stainless steels would be more susceptible to corrosion.

When potential increases in the anodic direction, first corrosion current density increases, resulting in ions that are corrosion products, which forms the protective coating oxide. The potential when the protective anodic oxide film is formed is called Flade potential, where the current density also decreases promptly. To keep the material passivated, potential must always be kept above the Flade potential; for that reason, either the anode must be polarized back to Flade potential with an oxidizer, or a material that has a high passivating property, such as stainless steel, must be chosen as the structural material.

## 5.3 Pourbaix Diagrams

The thermodynamic approach to corrosion process has been improved with potential (E) vs. pH diagrams developed by Marcel Pourbaix, which show areas similar to those in a phase diagram, where metals, ions, and oxides are both stable and unstable, that are separated from each other via the defining chemical (pH) and electrochemical (V) properties. Although they are prepared for 25°C, they can be used at higher temperatures such as 150°C with very little errors. Metal ion concentrations of $10^{-6}$ mol/l or more is assumed as the initial point of corrosion in these diagrams. Although the zones the metal where will be stable and the zones it will corrode if corrosion products are known via these diagrams, corrosion rate cannot be estimated.

In Pourbaix diagrams, if a boundary is parallel to the pH axis, it implies that the equilibrium does not involve $H^+$ or $OH^-$ ions, such as in the case of the following oxidation reaction of iron:

$$Fe \longrightarrow Fe^{2+} + 2e^- \qquad \text{(Eq. 75)}$$

On the other hand, boundaries parallel to the potential axis imply that equilibriums do not involve charge separations, as it is the case in the following reaction, where no species are oxidized or reduced:

$$Fe_2O_3 + 6H^+ \longrightarrow 2Fe^{3+} + 3H_2O \qquad \text{(Eq. 76)}$$

Potential-pH areas for equilibriums that involve either $H^+$ or $OH^-$ ions and charge separation have boundaries that are neither parallel to pH or potential axes, such as:

$$Fe_2O_3 + 6H^+ + 2e^- \longrightarrow 2Fe^{2+} + 3H_2O \qquad \text{(Eq. 77)}$$

One other piece of information that can be obtained from the pH-potential diagrams that assists in assessing the corrosion tendency is the solubility information. For instance, in the Pourbaix diagram of Fe-$H_2O$ system, areas where insoluble corrosion products such as $Fe_2O_3$ are stable indicate passivation areas, since $Fe_2O_3$ may slow down corrosion by covering the surface, while areas where corrosion products such as $Fe^{2+}$, $Fe^{3+}$, and $HFeO_2^-$ are stable as solutes define suitable conditions for corrosion.

Potential-pH diagrams have three major zones; the first is where iron metal remains in the metallic form not undergoing corrosion where $\Delta G > 0$, and thus corrosion cannot occur, which is entitled as the "immunity zone." It is also the principle of cathodic protection technique to establish potential conditions such that metal remains in the immunity zone. The second is where iron metal goes under corrosion, where $\Delta G < 0$, and thus corrosion can occur, which is entitled as the "corrosion zone," and the third is where corrosion products of iron prevent further corrosion, where $\Delta G < 0$ and corrosion is prevented from occurring, which is entitled "passivity zone." Another zone that is beyond the passive zone is sometimes referred as the "trans-passive" zone; however, it is basically another corrosion zone.

In essence, corrosion increases with an increase in potential from the equilibrium potential up to passivation potential, where corrosion rate decreases to one in thousand or less with a small increase in potential, and is called $i_{corr}$. Passivity is broken when potential is increased in the positive direction, leading to damage in the passive protective film, resulting mostly in pitting corrosion, and this region is called trans-passive region. Oxygen is needed for passivation to occur; thus, in stagnant neutral salt waters with low oxygen concentration, steel's passivity can be broken, or in other words, metal remains in the passive zone as long as the cathodic reaction rate is bigger than the anodic critical current density. The passivity zone is narrowed down with increasing acidity and temperature, resulting in an increase in the $i_{corr}$ value as well.

An iron (Fe) and water ($H_2O$) system is a very common system that well exemplifies Pourbaix diagrams. Ions of the Fe-$H_2O$ system are $Fe^{2+}$, $Fe^{3+}$, and $HFeO_2^-$, and solids are FeO, $Fe_2O_3$, $Fe_3O_4$, $Fe(OH)_2$, and $Fe(OH)_3$.

### 5.3.1 Immunity Region

At potentials more negative than –0.62 V, compared to standard hydrogen electrode (SHE), iron is thermodynamically stable, and is cathodically protected from corrosion; this zone is called as the immunity zone. Potential of –0.62 V corresponds to –0.850 V in saturated copper/copper sulfate reference electrode (CSE), which is the criterion reference point for cathodic protection of iron and steels.

### 5.3.2 Corrosion Regions

When potential is higher than –0.62 V and pH is lower than 9, iron corrodes, forming $Fe^{2+}$ and $Fe^{3+}$ ions, with $Fe^{2+}$ ions that are more stable at the lower regions and $Fe^{3+}$ ions that are more stable at the upper regions of the corrosion zone in the diagram. Another corrosion zone of iron in the diagram is when potential is between –0.8 V and –1.2 V and pH is higher than 13, according to the following reaction:

$$Fe + 2H_2O \longrightarrow HFeO_2^- + 3H+ + 2e^- \qquad \text{(Eq. 78)}$$

### 5.3.3 Passivity Region

The region above the two corrosion zones and the immunity zone is the passivity zone, where iron is anodically protected due to formation of $Fe_2O_3$ and $Fe_3O_4$ that result in the passivation of the metal surface, and protect the metal from corrosion.

## 5.4 Corrosion Equilibrium and Adsorptions

The corrosion process is essentially a surface phenomenon; thus, studying the physical chemistry of the surfaces reveals valuable information about the interactions of the metal atoms at the substrate surface with other molecules, e.g., the strength and the type of interaction with those of inhibitors. One such piece of information can be obtained from the Arrhenius equation, which is used to measured corrosion current with respect to temperature:

$$i_{corrosion} = k(e^{-E_a/RT}) \tag{14}$$

where $E_a$ is the activation energy. Since $i_{corr.}$ and T are measurable values, $E_a$ can be calculated from the slope of 1/T and $-\ln I_{corr.}$ graph. If $E_a$ is bigger in presence of the inhibitor, it would indicate that the inhibitor molecules are physically adsorbed on the metal surface and formed a protective layer. On the other hand, if $E_a$ is lower in presence of the inhibitor, it would indicate that the inhibitor molecules are chemically adsorbed on the metal surface.

Although adsorption is never at equilibrium on a surface that corrodes, it could be considered at equilibrium when the corrosion rate is sufficiently reduced and the Gibbs free energy for the adsorption is measured via the following formulas:

$$K_{ads} = 1/55.5e^{-\Delta G^\circ_{ads}/RT} \tag{15}$$

or

$$\Delta G^\circ_{ads} = -RT\ln(55.5K_{ads}) \tag{16}$$

If $\Delta G^\circ_{ads}$ comes out negative, it would mean that adsorption is spontaneous. Additionally, if it is a high numerical value, it would mean that there is strong interaction between the inhibitor molecules and metal surface. On the other hand, if the numerical value of $\Delta G^\circ_{ads}$ is less than 40 kJ/mol, then the adsorption is said to be physical. Additionally, adsorption enthalpy and its entropy are calculated via the following equations:

$$\log\left(\frac{Q}{1} - Q\right) = \log A + \log C - \frac{\Delta H_{ads}}{2.303RT} \tag{17}$$

$$\Delta G^\circ_{ads} = \Delta H_{ads} - T\Delta S_{ads} \tag{18}$$

If $\Delta H_{ads}$ is negative, the adsorption is exothermic, and if $\Delta S_{ads}$ is positive, along with negative $\Delta G^{\circ}_{ads}$, then the adsorption is spontaneous and is reaching equilibrium.

Provided that the corrosion rate is sufficiently reduced, thus assuming that adsorption is at equilibrium, inhibitor-metal interactions can be investigated via adsorption isotherms. For this reason, it is necessary to know the relation between the $C_{inhibitor}$, the concentration of the inhibitor, and Q, the ratio of the surface covered by the inhibitor, to be able to determine the isotherm that the adsorption fits. The following relation of $C_{inhibitor}$ and Q, for instance, is an indication of Freundlich isotherm:

$$\ln Q = \ln K_{ads} + n \ln C_{inhibitor} \tag{19}$$

where $K_{ads}$ is the equilibrium constant for the adsorption.

On the other hand, if $C_{inhibitor}$ vs. $C_{inhibitor}$ / Q graph is a linear one, then adsorption is said to be in correlation with the Langmuir isotherm, which is as follows:

$$C_{inhibitor} / Q = 1 / K_{ads} + C_{inhibitor} \tag{20}$$

If $K_{ads}$ is sufficiently large, it would imply that the inhibitor adsorbs on the metal surface strongly.

The potential of zero charges determined by Electrochemical Impedance Spectroscopy (EIS) is also helpful determining the adsorption behavior of inhibitors onto the metal surface. Charge of the metal surface is originated from the electrical field at the metal/solution interface, and can be measured by comparing the corrosion potential ($E_{corr}$) with potential of zero charges ($E_{PZC}$). Antropov named the difference between the two the relative corrosion potential ($E_r$), which is calculated as follows:

$$E_r = E_{corr} - E_{PZC} \tag{21}$$

If $E_r$ is negative, it would indicate a negatively charged surface, resulting in preference of adsorption of cations, and vice versa.

## 5.5 Concentration Corrosion Cells

Carriers of electricity current in the corrosion cells are electrons in the solid phase and ions in the electrolyte solution. Thus, total resistance of the corrosion cell is the sum of the external resistance of the solid phase and the internal resistance of the solution phase. Since electrons on the metal surface are connected to each other due to metallic conductivity, external resistance or electronic resistance is negligible. Direction of electricity flow is from cathode to anode in the solid phase, which is the opposite of direction of electron flow. Rate of corrosion or rate of dissolution of the anode is proportional with the corrosion current density based on Faraday's second principle. However, internal resistance sometimes becomes the primary factor preventing corrosion current to flow. Flow of current within the electrolyte requires the movement of both positively and negatively charged ions. Positively charged ions move to cathode, and negatively charged ions move to anode. Layers formed by corrosion products may prevent the movement of ions thus controlling the corrosion rate.

The sources of corrosion cells are the microscopic areas that show different electrochemical characteristics for both internal and external reasons. Examples of internal reasons are type of the metal and its structure. In highly crystalline materials, for instance, electrochemical properties of grain boundaries are different than of the bulk of the grains, since grain boundaries have higher energies for many reasons, such as the higher percentages of alloying elements at grain boundaries leading to the formation of microscopic corrosion cells, making the grain boundaries anode, resulting in intergranular corrosion. Examples of external reasons are different levels of oxygen contact with the corrosion deposits on the surface, differences in concentrations in general, and variations in temperature, which also lead to the formation of microscopic corrosion cells.

In a cell like Daniel cell that consists of a zinc electrode dipped into zinc sulfate and a copper electrode dipped into copper sulfate solution, the open circuit potential would equal to the difference between the reduction potentials of the two electrodes at zero current. When the electrodes are short circuited, both electrodes polarize and reach an equilibrium potential of $E_{corr}$, and the current passing the cell at $E_{corr}$ would be $i_{corr}$. Since electrodes have low polarization values in Daniel cell, the $i_{corr}$ is a large value. In real corrosion cells, $i_{corr}$ values are very low compared to the batteries. Such a case occurs when copper and steel metals are coupled in structures. For instance, pumps made of steel to pump sea water will corrode very fast due to the cathodic effect of bronze propellers unless propellers are covered with tin, which is more easily polarized compared to copper, thus leading to a reduction in corrosion potential and also in corrosion current. Similar incidents are observed in railways with electricity, in power lines carrying high voltage direct current, in the case of stray currents originated from welding machines, and in hot water systems, where pipes with diameters more than 50 mm are made of steel, and the ones with diameters less than 50 mm are made of copper.

Gibbs free energy of concentration cells can be calculated similarly as it is calculated for a galvanic corrosion cell in formulas 8 through 13. Gibbs free energy of the oxidation half-cell or the anode, where there is dissolution of metal resulting in an increases concentration of metal ions, is:

$$-\Delta G = nFE = RT\ \ln\left(c_1/c_2\right) \tag{22}$$

and Gibbs free energy of the reduction half-cell or the cathode, where there is reduction of oxygen in neutral and basic environments, is:

$$-\Delta G = nFE = RT\ \ln(Po_2/P^*o_2) \tag{23}$$

where n is the oxidation state of the metal ions, F is the Faraday constant that is 96500 coulomb, E is the potential difference, R is gas constant that is 8.314 Joule/K.mol, T is absolute temperature in Kelvin, $c_1$ is the higher metal ion concentration, $c_2$ is the lower metal ion concentration, $Po_2$ is the oxygen pressure at the cathode, and $P^*o_2$ is the oxygen pressure at the anode.

Concentration cells are formed due to differences in aeration in underground pipelines, and lead to pitting corrosion and stress corrosion cracking. At equilibrium, current ($i_0$) is equal to both the anodic and cathodic currents, $i_a$ and $i_c$.

$$i_0 = i_a = i_c \tag{24}$$

If equilibrium potential shifts in the direction of the cathode, the net current passing through the cell would be

$$i_0 = i_c - i_a \tag{25}$$

and if the shift is in the direction of the anode, then

$$i_0 = i_a - i_c \tag{26}$$

Thus, if $i_0$ is too large, such as in the case for platinum dipped in aqueous solutions, then hydrogen reacts instead of platinum, since the equilibrium current is too high for platinum and cannot be changed much, and this cell is called standard hydrogen electrode (SHE) and is accepted to have a potential of 0 at 298 K. On the contrary, if the equilibrium current of a cell is not too high, then it can be changed with the application of a small external current, or, in other words, it can be polarized easily. The relation with the applied external current and the potential change it causes is given in the Bulter-Volmer equation, and is very important to assess corrosion in corrosion studies.

$$i = i_0\{\exp\frac{(1-\beta)F\eta}{RT} - \exp\frac{(-\beta F\eta)}{RT}\} \tag{27}$$

## 5.6 Polarization

The potential difference between the cathode and anode is the driving force of any electrochemical process. However, the cathode's and anode's potentials are equalized when both electrodes are connected and a current passes through the cell, resulting in equalization of $E_c$ and $E_a$ at $E_{cor}$, which is a steady-state potential, where anodic current density equals to cathodic current density, leading to same anodic and cathodic reaction rates. That steady-state potential ($E_{corr}$) is called corrosion potential, and the corresponding current is called corrosion current ($i_{corr}$). Corrosion current mainly depends on the corrosion potential, diffusion of corrosive electrolytic species, temperature, and the ratio of anodic area to cathodic area. The diffusion of corrosive electrolytic species is the primary requirement to continue corrosion. The diffusion process is often slowed down due to deposition of products at the anode and cathode plates, and hence, corrosion rate decreases. With the increase in temperature, corrosion rate increases due to increase in diffusion rate. The corrosion current ($i_0$) is, as it is stated in formula 25, $i_0 = i_a - i_c$, where, $i_a$ and $i_c$ are the anodic and cathodic current densities, respectively. Thus, if the cathodic area is larger, anodic current density is significantly high, and hence corrosion rate increases as $i_a$ is much higher than $i_c$.

Cathodic protection aims to polarize the potential of the metal to be protected to the point of the anode's open circuit potential, making anodic currents zero, which is achieved by applying an external current to the metal in the cathodic direction.

Despite equilibrium potential providing good information about how the corrosion reaction is progressing, it does not provide any kinetic data about how fast the reaction is progressing. To measure the corrosion rate, an external current must be applied, and the change in the electrode potential must be measured. Polarization is the change in electrode potential due to this externally applied current, and polarization value is the difference between the equilibrium potential and the potential measured under the externally applied current. Butler-Volmer equation can be simplified for polarization values that are too small or large. First, if the polarization value ($\eta$) is bigger than 50 mV, then the second term in the Butler-Volmer equation can be omitted.

For $\eta > 50$ mV,

$$i = i_0\{\exp\frac{(1-\beta)F\eta}{RT}\} \tag{28}$$

for $\eta > -50$ mV,

$$i = i_0\{\exp\frac{(-\beta F\eta)}{RT}\} \tag{29}$$

and the natural logarithm of the first equation results in

$$\ln i = \ln i_0 + \frac{(1-\beta)F\eta}{RT} \tag{30}$$

and

$$\eta = \frac{-RT\ln i_0 + RT\ln i}{(1-\beta)F} \tag{31}$$

since $\eta$ and i are the only variables, formula 30 yields the following formula, which is called the Tafel equation:

$$\eta = a + b\ln i \tag{32}$$

Thus, in the case of high polarizations, the natural log of externally applied current and the polarization are directly proportional with one another.

Second, if the $|\eta| < 5$ mV then based on McLaurin series, exponents can be rewritten and only the first two terms can be considered:

$$i = i_0\{1 + \frac{(1-\beta)F\eta}{RT} - 1 + \frac{\beta F\eta}{RT}\} \tag{33}$$

leading to

$$i = i_0\frac{F\eta}{RT} \tag{34}$$

and

$$\eta = \frac{RT}{i_0 F}i \tag{35}$$

Thus, polarization changes depending on the current density. This method used for corrosion studies is called the linear polarization method. Information about polarization effects gives an idea about which reactions take place at the electrodes. There are several types of polarization:

### 5.6.1 Activation Polarization

Activation polarization occurs when corrosion reactions proceed at the metal-electrolyte interface, and the corrosion rate can only be measured through activation polarization.

### 5.6.2 Concentration Polarization

Concentration polarization occurs due to the concentration changes around the electrode in time. When current passes through the corrosion cell, some ions are consumed and their concentrations are reduced, leading to an increase in the electrode potential. Consumed ions must be replaced with new ions from the electrolyte, but this process is limited with diffusion rate. For instance, in spring waters and in seawater, the main cathodic corrosion reaction is oxygen reduction, and solubility of oxygen in the water is very low. Thus, in stagnant waters, oxygen concentration around the cathode is quickly reduced since it cannot be replenished that fast, leading to a quick cathodic polarization.

### 5.6.3 Ohmic Polarization

Ohmic polarization is due to the resistance in the circuit, which is also called IR ohmic potential reduction. Ohmic polarization is directly proportional with the current intensity. If current is stopped, then reduction in the potential in the amount of IR also disappears simultaneously, while other polarization effects decrease slowly.

## 5.7 Polarization Curves

When a current passes through a galvanic cell, cathodic potential shifts to more negative, while anodic potential shifts to more positive values, eventually reaching an equilibrium potential, where two potentials are the same. If external current is applied to such a corrosion cell in equilibrium, corrosion potential ($E_{corr}$) shifts to negative direction and the difference in potentials under applied current and without equals:

$$\eta = E_i - E_0 \tag{36}$$

The thermodynamically determined corrosion tendency concept reveals the conditions under which the metals tend to corrode; however, it does not yield any information about the rate of corrosion, which is more important, practically and economically. Thermodynamically for equal activities and under standard conditions, $\varepsilon_{(Cu/Zn)}$ was found as 1.126V; thus, if these two electrodes are connected through a voltmeter that has a high enough internal resistance, it will read a potential very close to 1.126V; however, if they are connected over a limited resistance, then a certain amount of current will pass through the corrosion cell, leading to a decrease in the potential value, which the voltmeter reads. These alternate values can be denoted with a prime as shown in the following equation:

$$\varepsilon' = i' (R + r) \tag{37}$$

where R is the external and r is the internal resistance of the solution or electrolyte, while ($\varepsilon - \varepsilon'$) is shared by both the anode and cathode electrode potentials, causing deviations from the potential values obtained when measured separately, resulting in both potential values approaching each other, thus given that additional potentials shared by anode and cathode are denoted with $\eta_{A'}$ and $\eta_{C'}$, respectively:

$$\varepsilon = i' (R + r) + \eta_{A'} + \eta_{C'} \tag{38}$$

or

$$\varepsilon - \varepsilon' = \eta_{A'} + \eta_{C'} \tag{39}$$

both $\eta_{A'}$ and $\eta_{C'}$ increase with increasing current ($i'$), while they disappear when there is no current. The deviation from the open circuit potentials of both anode and cathode is called polarization. In the case that external resistance is taken as negligible and 0, then the current will reach to its maximum value, while the value voltmeter will read:

$$\varepsilon'' = i''.\, r \tag{40}$$

In real environments, during corrosion of metals in aqueous solutions, external resistance becomes negligible, and leads to the micro anode and cathode electrodes to be in a short circuit situation through the metal resulting in polarization:

$$\varepsilon'' = i''.\, r + \eta_{A''} + \eta_{C''} \tag{41}$$

where i'' is the real corrosion current and ε'' ($E_{corr}$) is the corrosion potential that is between the anodic ($\varepsilon_{A''}$) and cathodic ($\varepsilon_{C''}$) potentials, which are very close to each other numerically due to polarization; thus, the polarization curve is linear and ±10mV within the $E_{corr}$, making it possible to establish linear equations such as the Stern and Geary technique, which works well for determination of the corrosion rate in acidic and neutral environments:

$$i_{corr} = 1/2.3\ (\beta_A \cdot \beta_c\ /\ \beta_A + \beta_c)\ (di/dE)_{i=0} \tag{42}$$

Thus, in the case of iron exposed to corrosion, if $E_{corr}$ is shifted by 10 mV either in the anodic or cathodic direction, resulting in a current of $0.9\times10^{-5}$ Amp/cm², and assuming that both $\beta_A$ and $\beta_c$ are 0.1, since theoretically available $\beta_A$ and $\beta_c$ values are usually between 0.06 V and 0.12 V, then $i_{corr}$ is found as:

$$i_{corr} = 1/2.3\ (0.1\times0.1/0.1+0.1)\ (di/dE)_{i=0} \tag{43}$$

and

$$(di/dE)_{i=0} = 0.9\times10^{-5}\ \text{Amp/cm}^2\ /0.01\text{V} = 0.9\times10^{-3}\ \text{Amp/cm}^2\text{V}_0 \tag{44}$$

Thus, $i_{corr}$ equals to $1.956 \times 10^{-5}$ Amp/cm$^2$ corresponding to around 50 mdd (mg weight loss per dm$^2$ and per day), which can be calculated using Faraday's second law:

Corrosion rate (mdd)
= c (mg/coulomb) . $i_{corr}$ (Amp/dm$^2$) . t (seconds/day) (45)

Rate of polarization is generally the most important factor determining the corrosion rate, since with increasing polarization, corrosion rate decreases, and thus one of the corrosion prevention techniques is to change the corrosive environment, increasing the polarization tendency. In formula 40, (i''. r) is usually negligible compared to ($\eta_{A''} + \eta_{C''}$). Also, it is usually the cathode among the two that shows stronger polarization tendency, and hence determines the corrosion rate.

In real corrosion reactions, two very common cathode reactions are the reductions of oxygen gas to hydroxide ions and of hydrogen ions to hydrogen gas. For instance, hydrogen ion reduction, which can occur for any metal dipped into an acidic solution, is shown via the following reaction:

$$2H^+ + 2e^- \longrightarrow H_2 \quad \text{(Eq. 79)}$$

and $\varepsilon_H = 0.0592pH$, while $\eta_H = \varepsilon - 0.0592pH$, where $\varepsilon$ is the observed potential due to passed current, and $\eta_H$ can be found from Tafel equation, which is:

$$\eta_H = \beta_H \log i/i_{0,H} \quad (46)$$

where i is the electron density per unit area on hydrogen electrode and of the metal, $\beta_H$ is Tafel slope, and $i_{0,H}$ is the exchange current density. Based on this equation, $\eta_H$ is highly dependent on the type of electrode material and its chemical composition. For instance, although the half electrode potentials of hydrogen potential are the same for platinum, iron, and lead, the $\eta_H$ values are very different. Among these metals, platinum has limited polarization tendency due to its high catalytic activity, while hydrogen reduction on metals that have low catalytic activity such as lead and mercury is very slow, and these metals correspondingly have high $\eta_H$ values. Thus, for pH < 4, where the cathodic reaction is the reduction of hydrogen ions:

$$\varepsilon_{corr} = \varepsilon_H + \eta_H = 0.0592pH + \beta_H \log i/i_{0,H} \quad (47)$$

where $\varepsilon_{corr}$ and pH are measurable and $\beta_H$ and $i_{0,H}$ constants are theoretically available as Volt and Amp/cm$^2$, respectively, leaving only i or $i_{corr}$ to be calculated as Amp/cm$^2$ as well, which can be converted to other units such as mdd, that is, mg weight loss per dm$^2$ and per day using Faraday's second law. In the case where the cathodic reaction is the oxygen reduction reaction:

$$O_2 + 2H_2O + 4e^- \longrightarrow 4OH^- \quad \text{(Eq. 80)}$$

$$\varepsilon_O = -1.23 + 0.0592pH \quad (48)$$

when $pO_2$ = 1 atm. and where ε is the observed potential due to passed current, and

$$\eta_O = \varepsilon + 1.23 - 0.0592pH \quad (49)$$

When both terms ($E_H$ = 0.0592pH and $E_O$ = −1.23 + 0.0592pH) are compared, it can be noticed that $E_O$ is 1.23V nobler. In other words, based on this comparison, the corrosion rate of a metal in aerated neutral conditions is supposed to be higher than the corrosion in de-aerated acidic environments; however, this is not the case, for several reasons. First, polarization tendency of oxygen is very high; secondly, dissolved oxygen concentration is usually very small in the real environments, limiting the corrosion rate; thirdly, oxygen's diffusion rate to the metal surfaces where corrosion reaction takes place is also very slow, and thus since oxygen cannot diffuse to the metal surfaces in the same rate that it is consumed, measured potentials come up very different from the open circuit potentials, which is referred to as *concentration polarization*. Concentrations of oxygen approaching to zero at the metal surface result in $\eta_O$ being a lot higher, which is also limited by the presence of other reactions taking place in the same environment.

If pH = 7, ($E_O = -1.23 + 0.0592pH$) becomes –0.84 V, which is very close to silver's reduction potential of $E_{Ag}$ = –0.80V, and thus it may be expected that silver can be slightly affected by corrosion in aerated conditions, while copper that has $E_{Cu}$ = –0.337V is affected even more, resulting in fast dissolution of copper ions. Despite the fact that zinc's and iron's open circuit potentials are different since oxygen electrode polarizes very strongly, the corrosion rates of zinc and iron are very similar. Another factor affecting corrosion rate is that if pH < 4, both hydrogen and oxygen reduction reactions take place, increasing the overall corrosion rate.

Corrosion rate can be measured as the weight loss due to corrosion per unit area and time. The salt fog chamber test is the a common method conducted in two different ways: static corrosion tests are performed under constant temperature and constant humidity, e.g., ASTM B117 and DIN 50021 tests, while rotational corrosion tests are performed under varying temperature, humidity and electrolyte environments for different durations, e.g., ASTM G 85, ASTM B 605 and DIN 50018. However, experiment duration is very long, and especially when the corrosion rates are very low, results are not reliable. Also, real conditions can never fully be replicated in laboratory environments. Additionally, if corrosion type is not uniform corrosion, weight loss values would be meaningless. Formation of layers on metal surface during the experiment may lead to similar problems in terms of reliability of weight loss tests to determine corrosion rate. Thus, engineers that are in the process of selecting the right materials for their project or the corrosion prevention technique will find the weight loss test too long, and they employ other techniques to determine the corrosion rate. As for the traditional electrochemical techniques, problems arise when an external current is applied to measure the corrosion, since this externally applied current causes a deviation in the real corrosion potential. Techniques such as electrochemical noise aim to eliminate such disadvantages of both weight loss tests and traditional electrochemical tests.

In most cases, since corrosion products cannot be carried away from the anodes and oxygen concentration cannot be replenished at the cathodes fast enough, corrosion rate does not increase after a certain limit. Thus, in such cases, corrosion rate can only be increased up to a certain limit with increased current. Increased potential causes other reactions at the electrodes as well. As a result, corrosion rate depends on the potential difference between the anode and the cathode, as well as polarization of the electrode reactions. Corrosion in electrodes that have low equilibrium current densities, which could be polarized easily with a low voltage, can be prevented effectively.

# 6

# Corrosion Prevention and Protection

In general, corrosion prevention methods' focus is the interface of the metal surface and solution, since that is where the corrosion mostly takes place. Appropriate metal selection, along with appropriate design, prevents most of the corrosion at metal/solution interface. If metal/solution interaction could be cut completely, corrosion could be completely taken under control. Major corrosion prevention methods are appropriate design, selection of the right material and modification of the composition of the alloy, modification of the environment into a non-corrosive one, use of inhibitors, use of metallic, organic or inorganic coatings, anodic and cathodic protection, etc.

Corrosion protective measures are specific to the nature of the material, its environment, and service conditions. Stainless steels that are perfected via lowering the carbon content are used in medical equipment, food processing, and chemical industries exposed to atmospheric and marine conditions. For example, stainless steels that are perfected via lowering the carbon content are used in many areas such as in medical equipments and largely in food processing and chemical industries; therefore, the coatings that are used to coat the stainless steels and other similar materials are exposed to various atmospheric and marine conditions. Among these coatings, metallic coatings are more suitable for acidic, atmospheric, and partially aqueous systems, while paints are preferred more in atmospheric conditions and aqueous solutions. Coatings such as epoxy, polyurethane, and chlorine-rubber polymeric paints can last 15 to 20 years even in extremely corrosive environments. Preparation of metal surface is the first step in the protective coating process. In this step, the metal surface should be properly cleaned by degreasing and desealing. Degreasing is carried out by exposure to trichloroethylene or other volatile organic solvents. Desealing is carried out by sand blasting technique or by acid leaching technique or by other chemical methods. Inhibitors are especially used if replacing the metal is not feasible, such as in closed cooling water systems. Anodic protection is based on passivation of a metal that can be passivated via polarizing the metal in the anodic direction, reducing the corrosion rate down to one in one thousand. Anodic protection is mostly implemented in water and acidic environments, such as in sulfuric acid tanks. Cathodic protection appears to be the most effective and economic method to prevent corrosion in general in underground, water, and underwater systems, and in specific in high-pressure natural gas and oil pipeline systems, pier bases, ships, water and petroleum storage tanks, containers carrying chemicals, heat exchangers, reinforced concrete steels, etc. In this chapter, several practical methods to prevent corrosion will be reviewed.

## 6.1 Proper Design

The corrosion of metals depends on the design of the equipment. Geometry of the structure should not allow accumulation of corrosion products. The design of a structure should be such that retention of moisture is as low as possible, because corrosion occurs in the presence of moisture. Design should allow for complete drainage in the case of a shutdown as well as easy washing. Water accumulation at upper levels of the structure, not being able to fully empty the lower levels especially the boilers, and uneven distribution of the potential when cathodic protection is applied are all risk factors for corrosion, and should be taken into consideration during the designing stage. For example, electrical boxes exposed to the atmosphere should be so designed that water does not collect at the top. Tanks and other containers should be designed such that the whole of the liquid can be drained off completely. Riveted joints should be avoided, and welded joints should be used, which prevents crevice corrosion and caustic embrittlement. Components that are suitable for accumulation of moisture with no removal systems other than the natural evaporation would lead to corrosion. Thus, components should be made with water removal options, and they should be placed with consideration to the gravity in such orientations, so moisture is not accumulated. On the other hand, even if the design does not allow water to be accumulated, moisture can form due to condensation if warm gases contact with cold metal surfaces. Thus, sufficient thermal insulation and/or ventilation are also important, especially at locations where condensation may occur. Then again, ventilation differences within the structure may lead to heterogeneous distribution of oxygen, which may lead to corrosion as well. Joint locations, welded areas, narrow gaps, etc. also may have insufficient ventilation, and with static liquids that are accumulated at these locations, local corrosion may begin. Thus, the design should, if possible, include provision for the exclusion of air. Additionally, while designing materials, galvanic coupling should be prevented, either by sufficiently isolating them or by not using materials together that are galvanically very different from one another. Another important issue to take into consideration is that the equipment and materials used to connect structural components, such as screws and welding components, must consist of more noble materials than of the structural components, and should be sealed so no liquids can pass through. Stresses caused due to high temperatures during welding and the prompt cooling afterwards, combined with corrosion, result in stress corrosion cracking. To limit such problems, reducing the time the welded location is exposed to high temperatures is required, and that depends on the structure and composition of the components to be welded, which can be considered during the initial design. Despite the associated problems, welding all the sites where two metals come into contact will reinforce the structure, along with closing the gaps where corrosive chemicals can pass through, thus help preventing corrosion.

Civil engineers and architects shall consider corrosion at the designing stage, and civil engineers shall try to implement corrosion prevention techniques, which could be much less expensive than repainting or repairing the structure. Even if the design is not preventative of corrosion, it is still very important for another reason, which is that if the design is not appropriate, even repairing the corrosion damage, e.g., repainting the structure or employing surface treatment techniques, may not be possible, and corrosion products accumulated at different locations within the structure may cause stress and cracks. To come up with a design that prevents corrosion, it is important to determine the purpose the structure will be used for, the corrosivity levels of the environment that the structure will be in, the planned lifetime of the structure, and whether periodical repairs are planned.

Simplifying the design usually helps with prevention of corrosion. Simplifying the shape in general that has fewer angles, corners, sides, and inner surfaces would help prevent corrosion in a less expensive and easier way. Thus, a profile manufactured as a single unit is better than one made with more than one component. Also, closed surfaces and round elements are better than sharp corners and open surfaces, namely, round profiles instead of L, T, and U shaped profiles. Additionally, structural design should allow the components to be reached for repainting and repairs; the gaps and locations should be convenient for re-treatments if necessary. Furthermore, based on the nature of the environment, the structure's foundation may be elevated by placing concrete underneath, especially if the ground is often wet. Consequently, designs should include the following precautions:

1. Structures should not allow accumulation of static waters within;
2. Locations that are designed for flowing of liquids should not have roughness and should be smooth with no cavities;
3. Shapes and designs that would make cleaning and painting of the surfaces difficult should be avoided;
4. Effects of atmospheric corrosion must be taken into account in the design, especially at sites where there is atmospheric pollution;
5. Materials used for thermal and electrical insulation should not absorb water;
6. Contact of metals and alloys must be prevented at the design stage to prevent galvanic corrosion; and
7. Finally, the project must be designed in a way that reduces the structural stress of the structure but does not lead to stress corrosion cracking and other stress related corrosion and mechanical failures.

## 6.2 Choice of Material

Perhaps the most common and easiest way of preventing corrosion is through the judicious selection of materials once the corrosion environment has been characterized. Here, cost may be a significant factor. It is not always economically feasible to employ the material that provides the optimum corrosion resistance; sometimes, either another alloy and/or some other measure must be used.

Materials must be chosen carefully based on the application in which they will be used, geographical location, physical and chemical characteristics of the environment, temperature and pressure conditions, other materials with which the material will come into contact, etc. While choosing materials, their corrosion resistance, easiness to produce, abundance, and expense should be taken into account. In this regard, noble metals are commonly used for surgical instruments and ornaments, as they are most immune to corrosion. In terms of effect of the environment, if the environment is dry, many metals like aluminum or stainless steel can be used absent any corrosion protection. On the other hand, in wet environments, inexpensive materials like mild steel can be used, provided that they have protective coatings. For conditions with high temperatures and pressures, it is better to increase the corrosion resistant nature of the metal rather than using a protective coating. Additionally, structural and galvanic differences of different materials that will be used in the same project are very important, since use of a single material in structures is usually not economical. The following factors detail the characteristics of the corrosive environment to consider when choosing a material:

### 6.2.1 Purity of the Chemicals in the Environment

Chemicals must be checked for impurities, since even small amounts of impurities may accelerate corrosion, e.g., trace amounts of vanadium or chromium in nitric acid accelerate corrosion. Sodium and vanadium often attack the protective oxide films, giving rise to the formation of low melting compounds like $V_2O_5$ (600°C – 900°C) and sodium sulfate, which leads to intercrystalline failures. Sometimes formation of carbides leads to changes in alloy composition, rendering the crack formation within the matrix, e.g., precipitation of chromium carbides in Ni-Cr alloys in a redox atmosphere followed by oxidation through the chromium-doped matrix, which results in "green decay" corrosion.

### 6.2.2 Electrolyte Concentrations

Dilute solution concentrations do not always impede corrosion, e.g., 10% $H_2SO_4$ solution is more corrosive than 90% $H_2SO_4$ solution.

### 6.2.3 Nature of the Electrolyte

Turbulence effects and presence of solid particles in the flowing liquid and also presence of stagnant areas on the path of the flowing liquid are among the factors to be considered when choosing the right structural material.

### 6.2.4 Effect of Corrosion Products

Corrosion products must also be taken into consideration, e.g., copper ions may diffuse into food from copper pots and lead ions may diffuse into water from lead pipes, which are both toxic.

### 6.2.5 Temperature Variations

Temperature increase in general increases corrosion, with exceptions.

### 6.2.6 Presence of Oxygen

The presence and sometimes absence of oxygen causes corrosion, e.g., 2% $H_2SO_4$ solution cause corrosion in austenitic stainless steels even in absence of oxygen, since in presence of oxygen steel will be passivated; however, in the case of copper/nickel alloys, the opposite is observed. Also, presence of oxygen encourages the fretting corrosion that mainly arises through the formation of oxide debris like red rust due to two closely fitting metal surfaces subjected to vibration. Fretting corrosion can be avoided by increasing humidity, reducing vibration and load, or using a bonded coating of $MoS_2$, etc.

### 6.2.7 Oxygen Concentration Cells

During designing stage, blind spots must be avoided so that oxygen concentration can be same throughout the structure, avoiding formation of corrosion cells due to concentration difference.

### 6.2.8 Interference Effects

Stray currents in the ground also cause corrosion; thus, the origin must be investigated and preventative measures must be taken.

In the light of aforementioned parameters, there are three major ways to employ metals in structural projects:

*i. As Pure Metals*

Pure metals have higher corrosion resistance. Even minute amount of impurities may lead to severe corrosion, e.g., 0.02% iron in aluminum decreases its corrosion resistance.

*ii. As Alloys*

Both corrosion resistance and strength of many metals can be improved by alloying, e.g., stainless steels contain 12%–18% chromium, which produces a coherent oxide film that protects the steel from further attack, allowing it to be used in manufacturing several instruments and equipment including turbine brackets, heat-resisting parts, etc. Stainless steels are susceptible to corrosion in some environments, and therefore are not always "stainless." Other alloying elements such as Al, Ni, Ti, Mo, etc. also form a protective oxide layer film, and tungsten (W), tantalum (Ta), niobium (Nb), titanium (Ti), etc. form protective carbides on the metal surface, while Al, Be, Si, etc. form stable oxides on copper surface, minimizing scaling in addition to protecting the underneath metal from corrosion. Among other examples are 0.1% cerium (Ce) content in Ni-Cr resistance wire, which increases its life ten times through the formation of interlocking between the oxide and the scale; traces of beryllium and calcium in Magnox alloy prevent burning; some alloying elements form low melting point oxides.

*iii. By Annealing*

Heat treatment like annealing helps to reduce internal stress and reduces corrosion.

## 6.3 Protective Coatings

Physical barriers to corrosion are applied on surfaces in the form of films and coatings. It is essential that the coating maintain a high degree of surface adhesion, which undoubtedly requires some pre-application surface treatment. In most cases, the coating must be virtually non-reactive in the corrosive environment and resistant to mechanical damage that exposes the bare metal to the corrosive environment.

Protection of the coating depends on its porosity or permeability, which are inversely proportional. General methods of coating are using organic or inorganic paints, electrolysis, hot dipping, passivation of the surface via formation of a protective molecular film of anodic oxides employing inhibitors, etc.

### 6.3.1 Protective Oxide Films and Passivation

Application of an external current or providing an oxidizing environment to increase the thickness of naturally occurring oxide films is one of the major measures to produce a more corrosion resistant surface, which is named "passivation." In other words, passivation is a phenomenon of converting an active surface of a metal into a passive one by forming a thin, non-porous, adherent, and highly protective film over the surface.

Some normally active metals and alloys lose their chemical reactivity and become extremely inert under particular environmental conditions, e.g., surfaces of aluminum, tin, chromium, iron, nickel, titanium, and many of their alloys rapidly get converted into oxides when exposed to the atmosphere or to the oxidizing environment. The non-porous natures of these oxide layers prevent further corrosion, and if damaged, they normally reform very rapidly. However, a change in the character of the environment, e.g., alteration in the concentration of the active corrosive species such as chloride and sulfate ions, causes a passivated material to revert to an active state, accelerating the film breakdown process, whereas chromates and phosphates promote the repairing action. Subsequent damage to a preexisting passive film could result in a substantial increase in corrosion rate, by as much as 100,000 times. Oxide films' corrosion resistance properties depend upon the properties of the film, such as:

a) Thickness and impermeability to media
b) Adherence to the base metal
c) Resistance to chemical attack
d) Mechanical strength
e) The ability to repair defects developed in the film

Metals that are susceptible to corrosion can be made passive by alloying them with one or more metals that are already passive and resist corrosion, e.g., iron is rendered passive by alloying it with any of the transition metals such as chromium, nickel, and molybdenum.

Dipping metals into solutions of chemicals or spraying such chemicals onto the metal surface to form an insulating coating as it is done with phosphates and chromates, to form chromate conversion coatings with the latter, is a common technique of passivation. Phosphate coatings have thicknesses that vary from 1 μm to 10 μm; they form a good base for the paints and reduce friction, especially during cold working. Metals such as aluminum, zinc, cadmium, and metals or coatings such as copper, silver, tin, or even phosphate coatings are dipped into chromic acid or chromate solutions to be further protected from corrosion. Chromates also form a good base for paints, e.g., zinc, cadmium, and aluminum do not require any further surface pre-treatment after coated with chromates. Chromate layer is usually 0.5 μm thick and has the ability of self-healing at damaged locations of the coating; however, it is toxic, hazardous, and carcinogenic, and thus harmful to the environment and public health.

Passivation of steel can be achieved by exposing to nitric acid; however, addition of chromium, nickel, molybdenum would still be needed in environments where chlorides are present, since these elements are resistant to the attack of chlorides. Aluminum and its alloys can be further oxidized by treating them as anodes in solutions of sulfuric acid, chromic acid, or oxalic acid. The produced aluminum oxide coating can be treated with water vapor or with boiling water or other solutions to seal it off, closing the pores, making it very insulating and protective. The thickness of this coating should be regularly 10 μm inside the buildings, 20 μm in atmospheric conditions, and 25 μm in corrosive atmospheric conditions, while the thickness varies between 20 μm and 40 μm when such coatings are colored.

### 6.3.2 Coatings with Metals, Alloys or Materials that are Conductors

Metals used to coat another metal are either nobler than the host metal or more active.

#### *i. Use of Nobler Metals in Metal Coatings*

Examples of nobler metals are gold or silver plating onto copper or tin, lead, chromium, or nickel-chromium coatings onto steel. In this type of coatings, when there is a coating failure such as pores, cracks, etc., the host metal corrodes underneath and may cause danger due to unnoticeable mechanical failures; thus, coating must be continuous and of high quality.

#### *ii. Use of More Active Metals in Metal Coatings*

In the use of more active metals to coat the host metal, such failures do not cause any problems, since the metal used for coating corrodes preferentially in any case. Surfaces of more active metals can also be oxidized to become a chromate or phosphate layer, making the coating a double layer coating with more resistance to corrosion.

Oxidation potential of ferrous ion to ferric ion is –0.77 V. Zinc, aluminum, tin, and lead are placed at higher positions in the electrochemical series, and hence are more susceptible to being attacked by oxygen, and act as anodes. Thus, underneath iron remains bright at the cost of the cathodic metal film on its surface. This is an example of sacrificial protection. There are several processes of application of metal coating on steel surface:

a) Hot dipping

In this process, the base metal is dipped into the pool of molten metal to be deposited and kept on the base metal for sufficient time. After that, the base metal is withdrawn from the molten metal bath and allowed to cool it to room temperature. A thin film of protective metal is tightly adhered to the surface of the base metal. Zn, Sn, and Pb coatings are usually carried out by this method. The process of Zn-coating is called "galvanizing" and the process of Sn-coating is called "tinning."

b) Electroplating

In this technique, the metal to be coated is dipped into the salt solution, which contains the metal ions to be deposited. The metal plate to be coated acts as cathode, which attracts the metal ions of the salt solution. Usually graphite acts as anode. When current is passed through the solution, metal ions are deposited as metals on the cathode plate, forming a protective film on the cathode plate. Several metals like Ni, Cd, Sn, Au, Cu, Cr, etc. may be deposited on the base metal by this technique.

c) *Metal Spraying*

In this technique, molten metal to be deposited is sprayed over the surface of the base metal, followed by drying. Thus, a thin film of coated metal is formed on the surface of the base metal. One of the important advantages of this technique is that a film of protective coating can be formed on any finished article, or any finished article can be coated with a protective film by this technique. Mostly Al and Zn coatings are carried out by this technique.

d) Metal Cladding

Sometimes the expensive metal or alloy is protected by covering it with a thick film of protective metal, either by the pressing technique or by the hot rolling method. The thick cover is often called clad, and the phenomenon is called cladding. The cladded material possesses both the strength of the alloy as well as corrosion resistance property of the clad. One important example is duralumin, containing 4% Cu, 0.5% Mg, and 0.5% Mn, with a small amount of Si and Fe. Cladding material is aluminum.

e) Cementation

In this method, base metal and powdered metal to be coated to the base metal are heated to high temperature and kept that temperature for long time, in order to allow the powdered metal to diffuse into the base metal. As a result of this, an alloy layer is formed on the base metal surface. Zn, Al, Si, Cr, etc. are frequently used to form protective alloy layers on the base metal.

*Aluminum coatings* are mainly used for protection of steel and duralumin in sulfurous atmospheres. In general, aluminum coatings can protect several metals like Cu, Ni, Pb, Cr, etc., but accelerate the corrosion of Mg. Aluminum coatings are resistant to carbonates, chromates, acetates, nitrates, and sulfates in the pH range of 6.4 to 7.2, but are readily attacked by dilute nitric acid ($HNO_3$), sulfuric acid ($H_2SO_4$), phosphoric acid ($H_3PO_4$), and hydrochloric acid (HCl) solutions. Aluminum alloys are also attacked by alkalis, and thus inhibitors such as $Na_2SiO_3$ and $Na_2Cr_2O_7$ are commonly used to protect machines and containers in the soap industry. In industrial atmospheres, the rate of attack to aluminum is in the region of 0 to 0.5 $mg/dm^2/day$, e.g., in the case of food products.

*Zinc coatings* are widely used in protection of steels in atmospheric conditions, against water, acids, etc. in marine boilers, propellers, and rudders. The passivation of zinc usually depends on the formation of adherent film on its surface. Mainly oxide films, chromate films, carbonate films, and hydroxide films appear as protective films on the zinc surface. However, longevity of the protective coatings mainly depends on the thickness of the film, which decreases at a very slow rate (0.0002 inch per year).

*Tin coatings* have some distinct advantages over the other metallic coatings, since they are physiologically inactive, they are not corroded by nutritional products in the absence of oxidizers, and iron is cathodic to tin under certain conditions. Tin is moderately resistant to acid solutions in presence of air, while in the absence of air, hydrogen, which develops on the surface, increases the resistance to the flow of current and stops corrosion. Tin coating is frequently used in containers carrying milk, while fruit juices have a more corrosive action on tin. Tin protects steel significantly from the attack of distilled water and atmosphere.

*Lead coatings'* corrosion resistance mainly depends on the formation of a protective layer, which is effective within the pH range of 3 to 11. Thus, lead is resistant to sulfurous, chromic, and phosphoric acids, and to the atmosphere. However, it is readily attacked by hydrochloric, hydrofluoric, nitric, and formic acids, as well as nitrate solutions.

### *iii. Use of Both More Active and Nobler Metals in Metal Coatings*

An example combining both types of coatings for maximum protection is the coating used for car bumpers, which has four layers that are copper that is in contact with the metal surface, semi-bright nickel, bright nickel, and micro porous chromate. Such a coating intentionally encounters corrosion at the bright nickel level due to the micro porous chromate layer for decorative purposes.

### *iv. Use of Conducting Polymers in Metal Coatings*

Recently, coatings of conductive polymers such as polyaniline, polypyrrole, polythiophen, and polyacetylene are starting to be used for metallic corrosion protection. It is proposed that conductive polymers provide anodic protection at the metal surface, allowing thicker iron oxide layers to form and also limiting the active sites for corrosion to take place via barrier act. Conductive polymers are convenient because they are economical, environmentally safe, their oxidized form is stable, and they can be electrochemically synthesized, and can be thus used for corrosion protection of iron, steel, zinc, aluminum, and other metals that can be oxidized over corrosion protective measures based on paint coatings consisting of phosphates and chromates. Inhibition efficiencies of conductive polymers are not just due to the presence of π electrons they have, leading to a strong adsorption on the metal surfaces not allowing corrosive chemicals to adsorb, but also their big molecular volumes covering most of the metal surface, leading to a physical barrier property. Inhibition efficiencies of these polymers are proportional with the number of monomers one polymer molecule has. Coating metal surfaces with conductive polymers is done via electropolymerization. Potential used for electropolymerization should be lower than the metal or alloy's dissolution potential to keep metal in the passivated zone in order to prevent corrosion. Single layer coatings are effective at preventing corrosion, while multiple layers are more effective. Polymers that have π-bonds and conjugated bonds are more conductive.

### 6.3.3 Coating with Inorganic Materials that are Insulators

Coating with inorganic materials that are insulators, e.g., concrete, glassy enamel, glass, or brick is another method of protective coating. Among these materials, glassy enamels are primarily made of frit, which is a composite alkali metal aluminaborosilicate combination. Glassy enamels protect the metals that they are coated with against acids, alkalis, water, abrasion, and erosion corrosion, and also up to 500°C, and can be formulated in the production stage to better resist one of these factors. However, they are not protective against hydrofluoric acid and sodium and potassium hydroxides that are in melted or hot solution form. They can also crack under mechanical and thermal stresses.

Industrial glasses containing primarily silicon dioxide are used to coat surfaces of storage tanks, reactors, water heaters, pipelines, and valves that are made of steel after their surfaces are cleaned, and structural stress is removed at high temperatures with thermal treatment. Glasses absorb negligible amount of water and are resistant towards all acids except hydrofluoric acid and concentrated phosphoric acid. Glasses are also resistant towards alkaline solutions at room temperature, but become less resistant with increasing temperature. A disadvantage of glass coatings is the difficulty to make repairs.

Concrete coatings passivate the metal surface due to high alkalinity, and are not permeable to corrosive chemicals. Concrete coat ings also help pipelines submerge under the sea and other water bodies due to their high weights.

### 6.3.4 Coating with Organic Materials that are Insulators

Coating with organic materials that are insulators in general, such as paints, plastics, and rubbers, is a common and effective method of protective coating widely used on oil tanks, steel structures, and pipelines. The organic coatings have two advantages: they do not allow air or moisture to reach the metallic surface, and the pigments present in the coating act as inhibitor and thus prevent corrosion. Paints consist of three major components: organic compounds that are mostly synthetic polymers that determine the chemical and mechanical properties of paint and are responsible of adherence to the surface, pigments that provide color and corrosion resistance, and solvents that dry up and solidify the paint. Pigments are in three categories based on their corrosion prevention properties: inert pigments, such as aluminum scales and micaceous iron oxide, protect paint physically against external factors such as sun light or UV light, humidity, etc.; inhibiting pigments, such as chromates and lead based pigments, passivate the metal surface or increase pH of the environment or deplete oxygen; and the most common cathodically protecting pigment is zinc dust. It is important to select the right paint for the type of the metal to be painted, cleaning and preparing the surface well, using the right method to paint, and painting under the right conditions such as temperature, humidity, pollution, etc., for the painting operation to be successful.

Coating of metallic surface by paints, varnishes, enamels, or lacquers provide a protective coating surface and protect the metal or alloy from corrosion. However, before application of such coatings, the surface of metal should be properly cleaned to remove grease, dust, sand, scale, etc., as these affect adherence. Additionally, the performance of paints or enamels as protective coating against corrosion depends on their application technique. A slight negligence at any stage of paint application can cause complete failure, as cracks may develop. The following steps are necessary in the application of organic coatings:

i. Surface preparation
ii. Sanding
iii. Priming
iv. Finishing
v. Filling

The paint coating at the final stage should be continuous so that no pores are formed in order to make it impervious to gas and water, and it should be chemically stable.

In some tropical regions, such as in the Middle East and Africa for example, high environment temperatures, as well as strong ultraviolet (UV) radiation, may have some negative effects on organic coatings' construction and application process, since solvents or other volatile components in coatings volatilize faster than in their surroundings, which causes generation of large numbers of bubbles on the surface and inner layers of the coating film, leading to a decrease in the coatings' adhesion and physical properties shortening service life of coatings greatly. Among these factors, UV is the most important reason for aging of coatings, particularly due to UV-B radiation in the range of 290 nm and 320 nm wavelengths, which causes decomposition of binders in coatings. Due to this UV aging effect, especially on binders of coatings, high bond energy compounds should be selected as coating binders to prolong the service life of topcoats. These organic compounds are compounds with F-C, Si-O, or H-O bonds in chain such as fluorocarbon resins, polysiloxanes, and acrylic polyurethanes. Another technique is to apply UV-absorbing varnish over topcoats as a thin layer coating, preventing UV ray from penetrating into underlying topcoats, rendering organic binders less susceptible to attack. Light stabilizers, including UV-absorbent and radical scavengers, are essential parts of UV-absorbing varnishes. Hence, a multilayer system, such as an epoxy primer and a weatherable acrylic polyurethane topcoat and an UV-absorbing varnish on the top, would be more efficient at preventing the damages of UV-radiation. Epoxy primer should be of a suitable epoxy resin and a curing agent with the addition of a high boiling point solvent, and using a high quality defoamer system, it can be applied on steel surfaces at high temperatures as well, along with a UV-absorbing varnish that is composed of hydroxyl acrylic resin, UV absorbent, solvents, and additives, which also prevent the UV penetration up to 95%, resulting in the coating to pass a 2000 hours of weather aging test.

## 6.4 Changing the Environmental Factors that Accelerate Corrosion

Sometimes the environment appears corrosive, and modification of the environment may be required to protect the material, reducing the corrosion rate. Lowering the fluid temperature and/or velocity usually produces a reduction in the rate at which corrosion occurs. Many times increasing or decreasing the concentration of some species in the solution will have a positive effect; for example, the metal may experience passivation. The environmental conditions provide assistance to corrosion, and hence, by changing these conditions, corrosion can be controlled. For example, if temperature is decreased, like all other chemical reactions, rate of corrosion also decrease. In few cases, the reverse sometimes happens; for example, boiling of fresh water or salt water decreases solubility of oxygen in it, and hence rate of corrosion in such condition decreases. The presence of gases like $CO_2$, $SO_2$, $NO_2$, etc. in the atmosphere also accelerates the rate of corrosion of the metals as these gases dissolve in water to form corresponding acids. Change in pH of the solution also affects corrosion because as pH decreases, the evolution of hydrogen gas can replace ionization of oxygen as the cathode reaction. Addition of corrosion inhibiting substances or retardants in small amounts also affects the rate of corrosion. The environment can be modified via several major ways to prevent corrosion.

### 6.4.1 Reducing the Corrosivity of the Solution

Chemical and electrochemical reactions taking place at the metal/solution interface take place via the ions or molecules adsorbed at the metal surface. These adsorbed ions or molecules change the metal's potential assisting its dissolution, thus preventing their adsorption using inhibitors that alternatively adsorb on the surface, or changing the metal's potential periodically, preventing the establishment of the conditions suitable for metal's dissolution, reduces corrosion rate. Other parameters of corrosion rate are temperature, pressure, flow rate, etc. Removal of the corrosive chemicals from the environment is the most effective method. While anodic reaction is the dissolution of the metal, cathodic reaction is either one or more of the reduction reactions of hydrogen, oxygen, and/or another compound. Hydrogen reduction reaction can be controlled by increasing the pH level, and if pH cannot be adjusted due to the requirements of the system, then inhibitors that increase the potential, preventing hydrogen's reduction, can be added to the system. This way, $H^+$ ions reaching the surface becomes difficult, reducing the rate of the anodic reaction as well. Removal of dissolved oxygen preventing oxygen's reduction can be done via heating, vacuum, using chemicals, or using catalysts leading to oxygen cleavage before entering into the system, causing corrosion. However, it is important to note that limiting presence of oxygen will also limit passivation of metals that can passivate. Additionally, presence of $CO_2$ also leads to corrosion via pH reduction, due to carbonic acid formation and also via carbonate precipitation. Prevention can be done via removal of gaseous $CO_2$ or muddy carbonate precipitates. In both removal of $O_2$ and $CO_2$, use of inhibitors also increase the conductivity of the environment, leading to corrosion thus carefully planned. Sometimes, removal of the corrosive chemical may not be possible due to the nature of the process, such as if salt solution must be used, then all corrosion prevention techniques must foresee the presence of chlorides in the environment. Adjustment of the metal/solution interface can be done by employing either active or passive methods. Passive methods employ coating or painting of the surface or passivating surface via formation of a thin molecular film using inhibitors, changing the metal's potential in the direction, making metal's dissolution difficult by increasing the potential required for dissolution. If this potential change is done via employing an external current rather than using inhibitors, then it is an active method as other cathodic and anodic protection methods.

*i. Deaeration*

The presence of increased amounts of oxygen is harmful, since it increases the corrosion rate. Deaeration aims at the removal of dissolved oxygen. Reducing agents are frequently used to remove oxygen from the surrounding medium, e.g., sodium sulfite ($Na_2SO_3$), hydrazine, etc.

$$2Na_2SO_3 + O_2 \longrightarrow 2Na_2SO_4 \qquad \text{(Eq. 81)}$$

$$N_2H_4 + O_2 \longrightarrow N_2 + 2H_2O \qquad \text{(Eq. 82)}$$

*ii. Dehumidification*

In this method, moisture from air is removed by lowering the relative humidity of surrounding air. This can be achieved by adding silica gel, which can absorb moisture preferentially on its surface.

*iii. Removal of Acids or Salts*

If the environment is sufficiently acidic, corrosion is prevented by treating the metal surface with lime. Further, salts of surrounding medium are removed by using ion-exchange resins.

### 6.4.2 Inhibitors

One of the ways to inhibit metal corrosion is by adding chemical inhibitors. Inhibitors are substances that, when added in relatively low concentrations to the environment, decrease its corrosiveness. The specific inhibitor to be used depends both on the metal or alloy and on the corrosive environment. Inhibitors are added to cleaning baths, steam boilers, refinery equipment, chemical operations, steam generators, cooling systems such as automobile radiators, etc. There are several mechanisms that may account for the effectiveness of inhibitors. Some inhibitors only cover the surface by attaching themselves to the corroding surface and interfering with the oxidation or the reduction reaction. In other words, they adsorb on the anodic and cathodic sites to prevent adsorption of corrosive species and dissolution of metal ions. Other inhibitors assist in the formation of a protective film covering the surface, or they passivate the already present protective coating, and some only reduce the activity of corrosive species at the surface by reacting with and virtually eliminating chemically active species in the solution such as dissolved oxygen. Anodic inhibitors react with the ions of the anode and produce insoluble precipitates assisting in the passivation process. The so formed precipitate is adsorbed on the anode metal, forming a protective film, thereby reducing corrosion. Addition of anodic inhibitors up to a certain limit increases corrosion rate, due to increase in conductivity of the solution, while after that limit, metal is passivated, and thus corrosion is limited. Examples of anodic inhibitors are alkalis, molybdates, phosphates, chromates, etc. Cathodic inhibitors interfere with cathodic reactions, which are of two types, depending on the environment. Hydrogen ion reduction in the acidic solutions can be controlled by slowing down the diffusion of $H^+$ ions through the cathode. This can be done by adding organic inhibitors like amines and pyridine. They adsorb over the cathodic metal surface and act as a protective layer. In neutral solutions, cathodic reaction is the formation of hydroxide ions via reduction of oxygen, which can be prevented by eliminating oxygen from the medium by adding some reducing agents such as $Na_2SO_3$ or via deaeration. Vapor phase inhibitors are organic inhibitors that readily sublime and form a protective layer on the metal surface, such as dicyclohexylammoniumnitrite. They are used in the protection of machinery and sophisticated equipment that is commonly sent by ships. The condensed inhibitor can be easily wiped off from the metal surface.

To be successful in corrosion prevention by adding inhibitors, especially in closed systems, attention must be paid to several important parameters, including type of the metal used, composition of the corrosive environment, pH of the solution, whether solution has access to air or not, presence of microorganisms, temperature of the solution, and structural designs.

Acidic solutions are used frequently in cleaning of mild steel, in pickling, and mild steel is used commonly in manufacturing of storage tanks and reactions containers. Usually, there is no oxide and hydroxide layers on metal surface in acidic conditions, which assists both corrosive chemicals and inhibitors in reaching the surface easily. Most acidic inhibitors are organic compounds and include hetero atoms of which the inhibition efficiency increases in the ascending order of $O<N<S<P$.

Depending on the electron density of the electron donating heteroatom, aromaticity, steric effects, length of alkyl chain, etc., inhibitors either adsorb on the surface sites where metal atoms dissolve, or on cathodic sites, or both. In acidic media, the inhibitor molecules are protonated, and they adsorb on the positively charged metal surface through previously adsorbed sulfates, for instance. The protective layer that the acidic inhibitors produce helps prevent corrosive chemicals from reaching the surface and increases the activation energy for the electrochemical reactions between the metal surface and the corrosive chemicals that reach the surface. When metal is exposed to HCl solution, there are three options:

1. If metal is relatively positively charged, then anions such as negatively charged $Cl^-$ are adsorbed on the surface, and inhibitor and water molecules that are protonized in the acidic medium attach to the metal surface through the $Cl^-$ ions.
2. If metal is relatively negatively charged, then inhibitor and water molecules that are protonized in the acidic medium adsorb on the metal surface directly.
3. If metal is relatively neutral, then no adsorption occurs.

In neutral or basic environments, inhibitors adsorb on the metal surface along with hydroxide ions. Hydroxide ions form precipitates with metal ions when solubility product of the metal hydroxide is reached, partially covering the surface. Inhibitors used assume a duty of filling the pores and holes of the metal hydroxide precipitates, and thus are called oxide phase inhibitors, such as phosphates that contribute to the precipitate film via the metal phosphates they form.

On the other hand, some inhibitors, such as chromates or nitrates, do not directly contribute to the precipitation film, but help the film to be more continuous and protective. Chromates, nitrites, benzoates, borates, and phosphates are effective inhibitors of corrosion of annealed steel, while chromates, nitrites, and phosphates are more effective of corrosion of cast iron; thus, in a system including both annealed steel and cast iron, benzoates and borates cannot be used.

Contaminants, oils and even protective oxide and hydroxide films harden adsorption of inhibitors on the metal surface; thus, surfaces must be cleaned appropriately before application of inhibitors. Also, the presence of corrosive species is important in determining the type of the inhibitor used. For instance, if chlorides are present, then the inhibitors should keep the metal's potential in the area immune to pitting corrosion or in the passivation area, which is usually the case, leading to such inhibitors to be named anodic inhibitors or passivators such as chromates and nitrites. Other ions in the environment such as sulfates, which are very corrosive chemicals, even can help reduce the pitting corrosion due to chlorides, when their concentration is twice of chlorides at pH = 2, 10 times at pH = 7, and again twice at pH = 12. The same can be done, for instance, at pH = 7 with chromates when their concentration is 7 times more than chlorides, and with nitrates when their concentration is 0.4 of chlorides, with perchlorates when their concentration is twice of chlorides, and with chlorates when their concentration is half of chlorides. Concentration parameter is important, such that in acidic conditions, concentration of the organic inhibitor is directly proportional with the surface coverage they provide, leading to higher corrosion inhibition efficiency, while concentration of the anodic inhibitors must be over a certain limit to passivate the surface and prevent corrosion, otherwise corrosion rate increases.

Inhibitors are economical when they are used in limited amounts. To keep the amount limited, it is important to pay attention to the type of the inhibitor used as well as the way they are implemented. It is also crucial to know the possible corrosion mechanisms in the absence of inhibitors and the prevention mechanisms in the presence of inhibitors for maximum efficiency. Inhibitor efficiency depends on the molecular structure and the presence of polar groups on the molecules.

However, efficiency of an inhibitor also depends on the characteristics of the environment it is employed, type and structure of the metal, temperature, concentration of the inhibitor, dipping time, etc. If inhibiting efficiency increases with temperature, activation energy decreases in presence of inhibitors, and vice versa. Also, independent of temperature, if activation energy increases in presence of inhibitors, it would mean that inhibitors are adsorbed on the surface via physisorption, while the opposite would imply adsorption via chemisorption.

Physisorption occurs due to the electrostatic interaction between the charged metal surface and the ions at the solution. If the metal surface is positively charged, then it would be easier for the anions to be adsorbed. However, cations can also attach to the surface via acid anions. On the other hand, chemisorption occurs via charge transfer or charge sharing between the metal surface and inhibitor molecules. Transfer of electrons occur between transition metals of the surface that have empty and low energy orbitals with that of organic molecules that have atoms possessing free electrons. In this regard, if chemisorptions resembles intra-molecular bonding such as ionic or covalent bonding, physisorption resembles intermolecular bonding such as dipole-dipole interactions or London forces. Inhibitors usually retard corrosion by adsorbing onto metal surfaces. If inhibition efficiency increases with increasing concentration of the inhibitor, the inhibitor is an adsorption inhibitor blocking the active sites on the metal surface. Corrosion prevention mechanisms can be studied via methods such as potentiodynamic polarization, linear polarization, and electrochemical impedance spectroscopy, which are all employed based on inhibitor concentration. Adsorption depends on the following criteria:

1. Nature and charged character of metal surface
2. Inhibitor's chemical structure
3. Charge distribution on the inhibitor molecule
4. Factors that cause corrosion

One way to inhibit corrosion is by using oils that adsorb on the metal surfaces. Such oils separate hard surfaces from one another, prevent dissolution and heat damages due to friction, seal surfaces off, and make power transfer easier. Adsorption of organic or inorganic inhibitors on metal surface depends on the functional groups on the inhibitor, steric effects, aromatic character of the inhibitor, if any, and its electronic structure in general. Heterocyclic organic compounds, including nitrogen or sulfur atoms, inhibit copper corrosion by forming chelates with copper ions or by forming a physical barrier. Inhibitors that show aromatic character and include heteroatoms that have free electron pairs to donate such as nitrogen, sulfur, and oxygen are found to be effective in prevention of acidic corrosion. Iron tends to form coordinated covalent bonds with heteroatoms that have free electron pairs. The adsorption on the metal surface occurs via the free electron pair of heteroatom, such as nitrogen and $\pi$ electrons of the aromatic group with that of metal surface. Hence, organic inhibitors are also called adsorption inhibitors, and are mostly used in acidic solutions. When diamine derivatives are used as inhibitors, as length of alkyl chain on the inhibitor increases, the inhibition efficiency increases as well, due to the surface area that the inhibitor molecule covers and blocks. According to the generally accepted theory, organic inhibitor molecules replace the water molecules that are already adsorbed on the metal surface:

$$\mathrm{Org(sol)} + x\mathrm{H_2O(ads)} \longleftrightarrow \mathrm{Org(ads)} + x\mathrm{H_2O(sol)} \quad \text{(Eq. 83)}$$

### 6.4.3 Eliminating Galvanic Action

If two metals have to be in contact, they should be selected so that their oxidation potentials are as near as possible. Further, the area of the cathode metal should be smaller than that of the anode, e.g., copper nuts and bolts on large steel plate. The corrosion can also be reduced by inserting an insulating material between the two metals.

## 6.5 Changing the Electrochemical Characteristic of the Metal Surface

The electrode potential of a metal can be easily changed by using electrochemical methods. As corrosion in aqueous conditions is an electrochemical process, cathodic or anodic protection provides a method to control corrosion. Out of the two protections, cathodic protection is more common and widely used in preventing corrosion of water heaters, underground tanks and pipes, and marine equipment.

### 6.5.1 Cathodic Protection

One of the most effective means of corrosion prevention is cathodic protection, which protects the metallic structure by converting it from anode to cathode externally, providing the electrons needed for cathodic reactions. Please refer to the latter chapters for details of cathodic protection, which is the main theme of this book.

### 6.5.2 Anodic Protection

Anodic protection is similar to cathodic protection in the sense that an external current is applied onto the metal; however, in principle, it is very different. External current is used for passivation of the metal, and thus this method can be applied only to metals that can passivate, of which there are only a few. Anodic protection is based on keeping metal's potential at anodic values compared to the corrosion potential, reducing corrosion risk. Thus, the metal is passivated by passing the carefully controlled current, and then a suitable potential is maintained. Parameters such as critical current density required for passivation, the passivation current, and the potential required for complete passivation need to be determined before anodic protection is applied. Applied potential and current must be controlled very carefully with the help of a reference electrode, since otherwise the metal can corrode very fast.

In practice, the positive pole of the DC current supply is connected to the storage tank, to which the anodic protection is applied, and the negative side is connected to the cathode dipped in the solution. Externally applied current increases metal's potential in the anodic direction until it reaches the passive region. To prevent the metal from moving back in the active region, applied potential is tried to be kept a few hundreds of mV above the Flade potential, that is, the minimum potential needed for the metal to passivate ($E_{pp}$), and an amplifier is added to the direct current unit, automatically maintaining the current to keep the metal passive. A very small amount of current is usually sufficient to keep a metal at the passive region, and thus costs associated with the applied current are very limited in anodic protection, in contrast with cathodic protection; however, cathodic protection can be applied to all metals. Another difference is that while anodic protection is more appropriate for solutions that are severely corrosive, cathodic protection is more suitable for environments and solutions that are slightly or mildly corrosive. While the anodic protection system's initial establishment cost is high, the cathodic protection system's maintenance costs are high. Potential is controlled in anodic protection, while either current or potential can be controlled in cathodic protection.

Anodic protection can be used for equipments made of mild steel in sulfuric acid, of stainless steel in phosphoric acid, of nickel in nitric acid, and of titanium in ammonia, organic acids, and caustic solutions. Thus, if it is known that anodic protection can be applied, sulfuric acid tanks and heat exchangers can be made of mild steel instead of stainless steels. Anodic protection cannot be applied in solutions containing chlorides and in hydrochloric acid solutions, since chloride ions prevent passivation.

If no protection is administered, a mild steel sulfuric acid tank would promptly corrode, resulting in a dissolved iron concentration of 5 ppm to 20 ppm daily, depending on the temperature of the acid and the size of the tank. If anodic protection can be applied, dissolved iron concentration increases only 1 ppm daily. This way, the tank's service life can be prolonged and the contamination of the sulfuric acid can be avoided.

Apart from applying an external current, anodic protection can be provided by administering a chemical method as well. Connecting the metal to be protected with a nobler metal that has high cathodic potential such as Pt or Pd or use of anodic inhibitors that can move the metal to the passive region are among such chemical methods. Trace amounts of Pt or Pd in the stainless steels or titanium would be sufficient to keep the substrate metal in the passive region.

The biggest problem faced in practice is the very high initial amounts of current needed for anodic protection, which is usually 100 times more than the current needed to keep the metal in the passive region or to maintain the system. To deal with this problem, either the metal is moved into the passive region very slowly, thus keeping the corrosion rate relatively very small, reducing the needed current or reducing the temperature of the solution, or adding an inhibitor to the solution that accelerates the passivation process. For instance, stainless steel tanks and containers are first treated with nitric acid to provide passivation, using the anodic inhibiting effect of nitrates. Another method is to gradually fill the tank with the solution, and thus by reducing the wet surface area in the container, gradually passivating the inner surface of the tank.

To reduce the already low current used in anodic protection even further, a current-cutter is used; thus, the current consumed can be reduced down to 1%, e.g., current can be applied for 6 seconds and not for the following 600 seconds. The current provided for 6 seconds is sufficient to anodically passivate the regions that corrode within the 600 seconds; therefore, many pieces of equipment can be anodically protected using the same transformer-rectifier unit.

# 7

# Cost of Corrosion

It has been estimated that between 3.5% and 5% of an industrialized nation's income or its Gross National Product (GNP) is spent on corrosion prevention and the maintenance or replacement of products lost or contaminated as a result of corrosion reactions, e.g., rusting of the automotive body panels, radiator, and exhaust components. The British Hoar committee prepared a corrosion cost report indicating that corrosion costs 3% of British Gross National Product (GNP), of which 23% can be prevented. Batelle Columbus Laboratories estimated the losses due to corrosion in U.S. equaling to 4.9%, while National Bureau of Standards (NBS), currently National Institute of Standard and Technology (NIST), found it as 4.2% and both with an error margin of ± 30%. Both studies revealed that a maximum of 45%, a minimum of 10%, and an average of 15% of the corrosion cost can be prevented. The difference between the two studies in terms of corrosion cost that can be prevented is 10 billion dollars, equaling to about 0.6% of the American Gross National Product (GNP). The reason for such a difference and wide percentage ranges is because it is not clear to what extent the corrosion in automobiles can be prevented. The cost of corrosion in the U.S. was considered to be about 3.5 to 4.5% of the country's gross national product, resulting in about 70 billion dollars of loss in 1976, which increased to 126 billion dollars in 1982 and 276 billion dollars in 1991. The percentage losses are considered to be even higher in underdeveloped or developing countries, where corrosion protection measures are not sufficiently implemented. However, in practice, it is generally accepted that only up to 30% of the corrosion loss can be prevented. Generally, costs associated by corrosion can be categorized as follows:

## 7.1 Corrosion Preventative Measures

Extra measures taken during designing stage of the projects, use of thicker metals for corrosion allowance, use of more expensive metals that are corrosion resistant, and use of paints, coatings, inhibitors, anodic protection, and cathodic protection are among the costs associated with corrosion.

## 7.2 Lost Production Due to Plants Going out of Service or Shutdowns

The cost associated with a natural gas or a water pipeline th at is out of service for a few days, or the cost of a one day halt in the electricity production of a thermal power plant, can be huge. The loss of prestige due to such interruptions should also be considered as corrosion damage.

## 7.3 Product Loss Due to Leakages

Product loss due to leakage of a storage tank at a petroleum or gas station, or the leakage of a pipeline due to corrosion, is considered as corrosion damage. In addition to the product loss, environmental pollution and fire hazard are other problems associated with the leakage of the product, especially if the product is a flammable one, as in the case of petroleum storage tanks leakages underneath gas stations. Product loss may occur until the corrosion damage is noticed, identified, and repaired.

## 7.4 Contamination of the Product

Corrosion products would lessen the quality of the products, especially in the case of nutritional products, medicine, and cosmetics. Also, if lead pipes are used, corrosion products of lead intoxicate the drinking water being transported.

## 7.5 Maintenance Costs

High maintenance costs such as repainting and replacement of corroded equipment are included in the corrosion costs.

## 7.6 Overprotective Measures

Overprotective measures to prevent corrosion are also considered as corrosion costs. An example is when an 8.2 mm thick pipe is used instead of a 6.3 mm thick pipe in a pipeline that is 20 cm wide and 360 km long, which causes the additional use of 3350 tons of steel. Inner cross-sectional area decreases with the increasing thickness of the pipe, leading to about 5% less fuel to be transported and more energy required to pump the fuel, which translates into higher costs per unit fuel being transported. Additionally, sometimes too expensive materials in excessive thickness are used as structural materials to prevent corrosion, which is another sort of financial loss if, in reality, such heavy corrosion is not the case.

As a result of all the aforementioned categories and more, a general loss of efficiency occurs, which may not have been estimated leading to further problems.

As a consequence, from an engineering perspective, along with finding a safe solution to corrosion problem, the solution must also be economical. Therefore, sometimes cheaper metals that have low corrosion resistances may be preferred over metals that have high resistivity. While expensive metals such as silver, titanium, and zirconium have rates less than 75 mm corrosion rate per year, moderately expensive copper, aluminum, and stainless steels have less than 100 mm per year, and cheap metals such as cast iron and mild steel have less than 225 mm per year. Corrosion rate here is mm reduction in thickness per year in ambient conditions. In corrosive conditions, these rates may go up to less than 250, 500, and 1500 mm per year, respectively; thus, metals cannot be used absent a corrosion prevention method.

The method of taxation has a big effect on industrial companies, since they tend to use materials with shorter service lives in order to be able to deduct them from taxes. Governmental companies especially prefer the income they can get in one year over the income they can get over the long term; thus, they do not tend to use cathodic protection, since the financial benefit of cathodic protection shows itself only in years.

# 8

# Cathodic Protection

One of the most effective means of corrosion prevention is cathodic protection that can be used for all types of corrosion and may, in some situations, completely stop corrosion. Cathodic protection techniques are widely implemented in the industry for metallic structures such as underground pipelines, ports, piers, ships, petroleum storage tanks, water storage tanks, etc. that are embedded in electrolytes such as water, soil, concrete, etc.

Cathodic protection was first tried on a military ship named *Samarang* by Sir Humpry Davy in 1824. The ship's body that was made of copper was protected using zinc anodes; however, as a result of a successful corrosion inhibition, copper ions have been removed from the system, which led to living organisms such as mosses to cover the ship's body, which in turn gave the impression that the cathodic protection attempt was unsuccessful. Cathodic protection was also used as early as 1836, by dipping iron sheets into molten zinc to protect war ships from corrosion. After about a century, cathodic protection was used again, this time for the protection of underground pipelines, and around the 1950s, it began to be commonly used for water and oil storage tanks, ships, dams, pier bases, reinforced concrete steel bars, etc.

Cathodic protection simply involves supplying, from an external source, electrons to the metal to be protected, making it a cathode. Normally, electrons are produced at the anode and flow to the cathode, where they are used at the cathodic reaction. If these electrons are provided externally, then the anodic reaction cannot produce anymore electrons, while the cathodic reaction rate increases, and the anodic reactions do not take place at the surface of the metal to be protected, but on the surface of another anode in the cathodic protection system.

It is not required to protect an entire metallic structure such as a long pipeline system with cathodic protection. However, for cathodic protection to be an effective corrosion prevention measure, the corroded regions of the pipeline that are to be protected should corrode. Whether protecting an entire metallic body or parts of it, cathodic protection is effectively implemented in following conditions:

1. Resistivity of the ground in which the reinforced concrete pipeline is embedded should be less than 1300 ohm.cm
2. Chloride content of the ground in which the reinforced concrete pipeline is embedded should be more than 500 ppm
3. Areas that have characteristics similar to deserts or swamps that have sands and silts with high permeability
4. Areas that have stray currents

## 8.1 Sacrificial Anode Cathodic Protection Systems

One cathodic protection technique is sacrificial anode cathodic protection that employs a galvanic couple; the metal to be protected is electrically connected to another metal that is more reactive in the particular environment. The latter experiences oxidation, and, upon giving up electrons, protects the first metal from corrosion. In other words, a more active metal is connected to the metal structure to be protected, so that all the corrosion is concentrated at the more active metal, and thus saving the metal structure from corrosion. For current to flow through this new galvanic cathodic protection cell, there must be sufficient potential difference between the anode and the cathode to overcome the circuit resistance. Current withdrawn from the galvanic anode depends on the anode's open circuit potential and the circuit resistance.

This method is used for the protection of seagoing vessels such as ships and boats. Sheets of zinc or magnesium are hung around the hull of the ship and, being anodic to iron, they get corroded. Zinc and magnesium are commonly used as such, because they lie at the anodic end of the galvanic series. Since they are sacrificed in the process of saving iron (anode), they are called sacrificial anodes. The corroded sacrificial anode is replaced by a fresh one when consumed completely. Among other important applications of sacrificial anodic protection are protection of underground cables and pipelines from soil corrosion, and prevention of the formation of rust water via magnesium sheets that are inserted into domestic water boilers.

The process of galvanizing is simply one in which a layer of zinc is applied to the surface of steel by hot dipping. In the atmosphere and most aqueous environments, zinc is anodic to, and will thus cathodically protect, the steel if there is any surface damage. Any corrosion of the zinc coating will proceed at an extremely slow rate, because the ratio of the anode-to-cathode surface area is quite large. Oxidation of sacrificial galvanic anode constitutes the anodic reaction, and mostly reduction of oxygen constitutes the cathodic reaction. As a result of these reactions, the anodic region may become acidic, and metal hydroxides can cover the sacrificial anode's surface, causing problems. Galvanic anodes' current capacities and current efficiencies are constant. Based on the cathodic area to be protected and the time the protection will be provided, the number of galvanic anodes and the weights of these anodes can be determined.

## 8.2 Impressed Current Cathodic Protection Systems

In the impressed current cathodic protection, the source of electrons is an impressed current from an external DC power source. Impressed current is applied in an opposite direction to nullify the corrosion current and to convert the corroding metal from anode to cathode. The negative terminal of the power source is connected to the structure to be protected. The other terminal is joined to an inert anode, often graphite, which is buried in the soil, and the high-conductivity backfill material provides good electrical contact between the anode and surrounding soil. A current path-exists between the cathode and anode through the intervening soil, completing the electrical circuit. Since in this method current from an external source is impressed on the system, it is called impressed current method, and this type of protection is usually given in the case of buried structures such as tanks and pipelines.

If alloys such as iron and aluminum alloys that are not inert are used instead of noble platinum and palladium metals or inert graphite, they would dissolve promptly. Thus, cathodic polarization of the original anode is achieved by employing inert anodes making up the negative pole of the cell, providing the needed current externally to the system. This way, anodic reaction is changed from dissolution of any metal to gas evolution reactions, such as oxygen evolution in freshwaters and underground leading to acidic environments, chlorine evolution in seawater sometimes along with oxygen evolution, or carbon monoxide and carbon dioxide evolution in the case of graphite or coke dust anodic beds:

$$H_2O \longrightarrow \tfrac{1}{2}\,O_2 + 2H^+ + 2e^- \qquad \text{(Eq. 84)}$$

In the seawater or in waters that have high concentration of chlorides, chlorine gas evolves at the anode, either alone or along with evolution of oxygen:

$$2Cl^- \longrightarrow Cl_2 + 2e^- \qquad \text{(Eq. 85)}$$

and chlorine reacts with water, forming hypochlorous acid, which is a weak acid, and thus does not reduce the pH much:

$$Cl_2 + 2H_2O \longrightarrow 2HClO \quad \text{(Eq. 86)}$$

The fact of which of the two gases will be evolved at the anode between chlorine and oxygen gases is based on the chloride concentrations and over potential values of these gases at the anode. Standard electrode potential of chlorine is –1.36V and for oxygen it is –0.40V, while in the seawater, chlorine's dissociation potential is –1.39V and oxygen's is –0.81V; however, due to oxygen's high over potential value, which is 0.70V on the surface of steel substrate at $1mA/cm^2$ current density, only chlorine gas evolves at the anode for current densities that are not very high.

Usually, galvanic anodes used in terrains include coke dust as the anodic material, which oxidizes to carbon monoxide and carbon dioxide gases at the anode. Coke dust must be wet for the following reactions to occur, and as a result of these reactions, anode becomes acidic:

$$C + H_2O \longrightarrow CO + 2H^+ + 2e^- \quad \text{(Eq. 87)}$$

$$C + H_2O \longrightarrow CO_2 + 4H^+ + 4e^- \quad \text{(Eq. 88)}$$

It is important that the anodes used for both sacrificial anode and impressed current cathodic protection systems are economical, since almost half of the initial cost of cathodic protection systems is spent for the anodes. Secondly, current withdrawn from the unit surface area of the anode must be as high as possible, and the anodic resistance should not increase in time. Weight loss for the anode per current withdrawn in a year must be as small as possible in order for them to produce currents in planned intensities and durations.

Consequently, service life of inert anodes in this technique is expected to be long unless chemical reaction with other species and polarization occurs; however, inert anodes may be subject to weight loss due to passing current even of small amounts.

## 8.3 Cathodic Protection Current Need

Estimated values for cathodic protection current needs are established in the literature based on the type of the metal, surface area of the metal to be protected, type and the quality of the coating, as well as on the corrosivity of the environment that the system is within, e.g., environment's resistivity, pH, dissolved $O_2$, etc. However, usually these current values are given in a wide range, and while the lower limit of the range may not be safe, the upper limit of the range may not be economical as well. Secondly, coating quality is considered only based on coating resistance and thickness, but not based on commonly encountered human errors such as damages on the coating during transportation, storage, installation of the pipes, and from insulating the welded regions.

Corrosivity of soils, the earth, or the terrains in general is related to their resistivities, their pH, and redox potentials, which were already discussed in detail in the section 4.1.3, "Underground or Soil Corrosion."

Theoretically, the minimum needed current to cathodically protect a metallic system is calculated from the polarization curves of the anode and cathode reactions. For instance, steel's corrosion in an acidic environment consists of the cathodic hydrogen evolution reaction and the anodic iron dissolution reaction producing iron cations. At equilibrium of a corroding metallic system such as a corroding steel system, the potential would be $E_{corr}$ and the corrosion rate, $i_{corr}$, would be the point corresponding to the intersection of the anodic and cathodic polarization curves when graphed, e.g., $E_{corr}$ is –250 mV and $i_{corr}$ is $10^3$ μA/cm$^2$ (1mA/cm$^2$). If cathodic current is applied to this corroding steel system, shifting the potential in the negative direction down to –370 mV from –250 mV, the corrosion rate would be reduced down to one in thousand of the initial value to 1μA/cm$^2$, and the current intensity that is needed to do that would be $10^4$ μA/cm$^2$. In other words, current needed to protect a bare steel surface of 1 m$^2$ in an acidic environment would be 100 A. However, cathodic protection with such a high current intensity would not be economical, and thus, in acidic solutions, anodic protection is usually preferred over cathodic protection. However, if cathodic protection is a must, then at least the metal surface can be well coated to reduce the intensity of the needed current.

Regardless of the cathodic protection method used, the aim is to keep, e.g., iron materials' potential below –0.85V compared to $Cu/CuSO_4$ (CSE), and –0.55V compared to SHE. If there are sulfate reducing bacteria (SRB) in the environment, then the upper limit becomes –0.95V instead of –0.85V compared to CSE. Potentials more negative than these values bring about the problems of hydrogen damages, and are thus risky. Thus, not to take such risks, it is sufficient to reduce the corrosion rate to a certain level, and complete prevention is not attempted. For example, when metal's potential is shifted 100mV in the negative direction, corrosion rate is reduced by 90%, while a shift of 200mV leads to a 99% reduction in the corrosion rate. Generally, there is no need to reduce the rate of corrosion more than 99%. The current needed to provide such potential changes depends on the characteristics of the system. For a coated pipeline that is 240 km long, the total needed current can be 2 A, compared to a bare pipeline system of the same length that requires 1000 A.

Therefore, controlling the potential is usually the preferred method instead of adjusting the current, since too many variables are involved. However, determination of the total cathodic protection current is required in sacrificial anode cathodic protection systems to determine the number, mass, and the size of the anodes that will be installed.

Based on the environment and the quality of the coating, cathodic protection current needed per unit area usually falls into the following ranges in the case of cathodic protection of steel:

- for bare steel structures in fluidic, wavy, rough seawater, 100 to 160 $mA/m^2$;
- for bare steel structures in stagnant seawater, 55 to 85 $mA/m^2$;
- for bare steel structures embedded in the seabed mud, 20 to 30 $mA/m^2$;

- for bare steel structures in humid soils and underground, 10 to 20 mA/m$^2$;
- for weakly coated steel structures in water or underground, 1 to 2 mA/m$^2$;
- for well coated steel structures in water or underground, 0.05 mA/m$^2$;
- for polyethylene coated steel structures in water or underground, 0.005 mA/m$^2$.

These values are the initial currents needed for cathodic protection, since after polarization of the metal surface, the needed current decreases. However, coating resistance also decreases in time, leading to an increase in the current need, which may balance out the polarization effect.

## 8.4 Effect of Coatings on Cathodic Protection

Coating the metal that is to be cathodically protected reduces the amount of current needed for cathodic protection. This is possible since by coating the metal, diffusion rate of oxygen to the cathodic regions on the metal is reduced, and thus less current would be needed to polarize the metal cathodically. Coating the metal reduces both the initial establishment and operational costs of cathodic protection systems. Therefore, cathodic protection and coatings are used together.

## 8.5 Effect of Passivation on Cathodic Protection

Cathode reactions take place on the replacement anodes under the cathodic protection current in the form hydrogen and oxygen evolutions. These reactions produce hydroxide ions, which precipitate on the substrate surface along with the hardness of water, $Ca^{2+}$, and $Mg^{2+}$ ions. Similar to the effect of coatings on cathodic protection, these precipitates of calcium and magnesium hydroxides prevent diffusion of oxygen to the cathodic regions on the metal, resulting in the cathodic protection to proceed, with a smaller limit current reducing the cathodic protection current needed to polarize the metal. This explains the cathodic protection of the bare steel structures in seawater, since instead of the coating, the hardness of seawater and corrosion products form a protective layer on the metal surface, which acts similar to a coating. Therefore, although the applied cathodic protection current is initially very high, it decreases down to one-fourth of the initial amount just in a few months.

## 8.6 Automated Cathodic Protection Systems

Varying circumstances, such as changes in the underground water levels, changes in flow rates, etc., change the current needed for cathodic protection, and therefore, it has to be adjusted periodically. An automated transformer-rectifier (T/R) unit can provide this adjustment with a reference electrode placed underground along with a magnetic amplifier unit. When the potential measured by the reference electrode is below a certain limit, applied current is increased automatically, and vice versa.

## 8.7 Cathodic Protection Criteria

For steel structures to be cathodically protected, their potential must be sufficiently polarized. The following criteria are commonly used to check whether there is sufficient polarization:

### 8.7.1 –850 mV Criterion

Based on this criterion, potential of the steel structure that is protected under the externally applied current must be at least –850 mV or more negative compared to the potential of the saturated copper/copper sulfate (CSE) reference electrode.

If the terrain is anaerobic, containing sulfate reducing bacteria (SRB), for instance, then this potential should be –950 mV or less. Pipeline/ground potential must be measured at the "off" position after current is applied for a sufficient amount of time that is at least four hours, and also at the "on" position while the current is still being applied. One problem is that reference electrode cannot be placed directly on the underground pipeline systems, and thus the measured potential must be up to 200 mV to 300 mV higher in the negative direction, especially in grounds with high resistivity, to meet the –850 mV criterion making up for the potential difference. This difference depends on the distance of the reference electrode from the pipeline and the terrain's resistivity. For instance, for a terrain that has a resistivity (ρ) of 5000 ohm.cm and for a distance of 50 cm between the reference electrode and the pipeline, the measured potential should be –1.2 V at the surface; while for a distance of 100 cm, it should be –1.5 V.

### 8.7.2 300 mV Potential Shift Criterion

Based on this criterion, when cathodic protection current is applied on the steel structure to be protected, the measured potential should be shifted 300 mV in the negative direction compared to the static potential of the system, which was measured before the cathodic protection current was applied. However, since the system's potential is measured under the applied current, an ohmic potential reduction of IR must be taken into consideration. The difference of this criterion than the –850 mV criterion is that potentials less than –850 mV may be acceptable if the initial static potential was less than –550 mV, since then a 300 mV shift would result in a potential that was less than –850 mV.

### 8.7.3 100 mV Polarization Shift Criterion

In addition to the 300 mV potential difference criterion between the static potential and the potential measured under cathodic protection current, a 100 mV polarization shift should also be there between the initial static potential and the static potential measured after at least 4 hours of application of cathodic protection current and stopped for the purpose of measurement. Since the measurements are performed both in the "off" positions when no current is applied, no ohmic potential reduction of IR is observed, and thus, theoretically, this criterion is the most reliable criterion.

### 8.7.4 Tafel Region Starting Point

Potential values where cathodic protection is realized are graphed versus log (i), and the potential that corresponds to the Tafel region starting point is determined. Although this criterion has a very strong scientific basis, it is hard to apply in real life conditions in the field.

## 8.8 Reliability of Cathodic Protection Criteria

In practice, –850 mV criterion is usually preferred, since it is easy to perform. However, this criterion is reliable only at pH values close to neutral pH conditions and at low temperatures, and also IR ohmic potential reduction must be taken into consideration. –100 mV polarization shift criterion is good due to not including an ohmic IR reduction, while Tafel region starting point criterion indicates both the minimum potential and also the minimum needed current for cathodic protection, since E is graphed vs. log (i).

Consequently, cathodic protection criteria are usually reliable if the ohmic IR reduction is at reasonable values. For instance, for a steel pipeline that has a static potential of –500 mV compared to copper/copper sulfate (CSE) reference electrode, applied cathodic protection current results in the pipeline/terrain potential to go down to –750 mV at first, and then down to –850 mV after realization of the polarization meeting the first criterion. The second criterion is also met, since the difference between the "on" and "off" potentials is –850 mV – (–500 mV) = 350 mV. After the current has been interrupted, the potential is measured as –700 mV, and –700 mV – (–500 mV) = 200 mV, which is more than 100 mV, and thus the third criterion that is the polarization shift criterion is also met. –700 mV is 150 mV more than –850 mV, and thus considering the polarization shift values, this indicates there was about IR = 100 mV ohmic decrease in potential; therefore, as long as the IR ohmic decrease in potential can be kept around values of 100 mV, then any of these criteria can be reliable, with the first criterion being more preferable due to its easiness to determine; however, for larger IR ohmic potential reductions, 100 mV polarization shift criterion remains as the only reliable option.

## 8.9 Interference Effects of Cathodic Protection Systems

Generally, at locations around the anodic bed where the current is applied, a positive increase in the terrain potential occurs and other pipeline systems and metallic structures in the area are affected. This increase depends upon the terrain resistivity, current withdrawn from the anode, and the distance from the anodic bed, and is calculated according to formula 49.

The main types of interference effects that cathodic protection systems impose on the foreign metallic structures that are placed near to the cathodic protection systems are anodic and cathodic interference effects, while interference effects that are due to high voltage transmission lines, railway vehicles, welding operations at a ship, insulated flange interference, and ship-pier interference are also common. Regardless of the type, interference effects may cause severe interference corrosion due to stray currents if preventative measures are not taken.

### 8.9.1 Anodic Interference

Anodic interference is caused by the stray currents originated from the anodic bed of the cathodic protection system escaping to the surrounding metallic structures, where they cause corrosion not at the locations they enter the metallic structures, but at locations they leave the metallic structures, which have the lowest resistances.

In case of underground pipeline systems, a potential gradient in the positive direction is formed around the anodic bed when cathodic protection current is applied from an anodic bed that is at a certain distance from the pipeline, while a reduction in the potential in the negative direction is observed in areas nearby the pipeline. Going from the anode to cathode and nearby the anode, the potential decrease can be between 10 V to 50 V, depending on the produced current capacity of the anodic bed, while the potential decrease is negligible in the areas in between the anode and the cathode and farther away from both; lastly, the potential decrease is between 1 V and 2 V just around the cathode in a small region, depending on the quality of the coating of the pipeline. These regions of potentials around the anode and the cathode may cause interference on other structures that are nearby, leading to interference corrosion.

The ground's potential is increased in the positive direction when a high potential current is applied from the transformer-rectifier (T/R) unit of the impressed cathodic protection system to the anodic bed. This increase is directly proportional with the current dispersed from the anode to the ground and also with the ground's resistivity. Potential difference created at the surface is measured via the following formula:

$$\Delta E = \frac{i\rho}{2\pi r} \tag{50}$$

For instance, potential difference created at the surface, in a terrain that has a resistivity ($\rho$) of 3000 ohm.cm and for a cathodic protection current ($i$) of 10 Amperes, is 477 mV at a distance of $r$= 100 m away from the anodic bed. Metallic structures in the surroundings are affected by this positive potential difference, especially if they are bare or weakly coated, exposing them to the stray currents, resulting in anodic interference corrosion. The locations that the current enters the foreign metallic structure become the cathode, and the locations that current leaves the structure become the anode. If the location that the current leaves the structure is a narrow area, the corrosion can be severe. Interference corrosion also depends on the location of the anodic bed and its geometry. The following measures can be taken to prevent anodic interference corrosion:

1. During the designing stage at the project phase, anodic beds should be planned to be installed at least 70 m away from the surrounding metallic structures.
2. Parts of the foreign metallic body that remain in the anodic interference field are coated very well.
3. If interference corrosion is encountered during the operation stage, then galvanic anodes should be connected to the critical locations of the foreign metallic structure, e.g., another pipeline system, and thus the current that enters the structure can be discharged through these anodes.
4. Profile pipes or metal plates are placed on both sides of the part of the foreign pipeline that remain in the anodic interference region and are connected to the anodic bed's negative pole, acting as a shield for the foreign pipeline system.
5. A deep well anodic bed can be used instead of a shallow well anodic bed. This way, the center of the anodic potential field would be a spherical region that is 20 m to 30 m underground, substantially reducing the interference effect on the metallic structures at the surface.

### 8.9.2 Cathodic Interference

Cathodic interference is caused by the cylindrical negative potential reduction around the pipeline to be protected, resulting in stray currents originating from the surrounding metallic structures to enter the cathodically protected pipeline system, resulting in severe corrosion at locations where the stray currents leave the surrounding metallic structures, especially if these metallic structures are bare metals or weakly coated.

Interference corrosion at the foreign metallic structure occurs due to being in the region of negative potential decrease of the pipeline that is cathodically protected. In other words, interference corrosion occurs at the other metallic structure, for instance, at another pipeline that is not cathodically protected, since it enters to a region of more negative potential than the region it is coming from. This difference in the potentials of two regions that the other pipeline is passing through results in the part of the pipeline that is in the more negative potential area to become the anode and corrode. This type of interference corrosion increases with increasing cathodic protection current.

The length of region that is under cathodic interference effect is measured by potential measurements on the foreign metallic structure, e.g., another pipeline. Measures that can be taken to prevent cathodic interference corrosion in the foreign metallic structure, which mainly aim to reduce the potential difference, are as follows:

1. Pipeline that is cathodically protected and the foreign pipeline or metallic structure are connected electronically via a metallic connection where they are physically closest to one another, and thus this way the stray currents escaping through the terrain are carried via a conductor. Stray current intensity must be just enough to meet the potential difference, and thus a resistance of appropriate size is installed between the two metallic structures based on the metallic electronic connection.
2. A metal plate or a scrap metal is placed, in between the pipeline that is cathodically protected and the other pipeline, at a location where they are the closest to one another and connected to the other pipeline. Consequently, interference corrosion takes place on this scrap, junk metal instead of the other pipeline. Theoretically, the farther away the scrap metal is from the other pipeline, the more it can protect the pipeline from interference corrosion;

however, in practice, this would also lead to an increase in the amount of the produced cathodic current; thus, scrap, junk metals, or metal plates would be placed along the other pipeline and just somewhere in between the two pipelines.
3. The parts of the other pipeline that remain in the potential decrease region of the cathodically protected pipeline must be well coated.
4. The regions of the unprotected foreign pipeline or metallic structure that remain in the potential decrease region of the cathodically protected pipeline are connected to an appropriate number of galvanic anodes. This way, stray currents enter the foreign pipeline through these galvanic anodes. These anodes do not have to be high potential anodes; usually, zinc anodes are sufficient. Additionally, these anodes do not need to be placed in between the two pipelines; it would be enough to place them as near as possible to the foreign pipeline only.

### 8.9.3 Specific Interference Cases

Other than anodic and cathodic interferences, there are also other types of mixed interferences:

#### *i. Interference Effect of Railway Vehicles*

Often stray currents escape from railway vehicles that work with direct current to the surrounding metallic structures. Railway vehicles work with very high voltages, and currents of thousands of amperes are withdrawn during their motion. The positive pole of the direct current is connected to the cable feeding the train, while the negative pole is connected to the rails. As the train moves, current passes on to the rails and goes back to the station. At the locations that the train is passing through, terrain potential increases in the positive direction, and stray currents escape to the terrain from the rails. These stray currents may go back to the station over the pipeline if a pipeline system is present in the surroundings. As the train moves, the locations where the stray currents enter the pipeline change. In fact, the locations where the stray currents enter the pipeline are not important. The location where the currents leave the pipeline is very important, since that is where corrosion occurs. When the train is at the station, rails and terrain have more negative potential than normal; however, as the train gets away from the station, both the rail's and the terrain's potentials increase in the positive direction.

Currents enter the pipeline easily if the pipeline coating is damaged and of low quality. The presence of insulated flanges on the pipeline, and the fact that the path of pipeline gets away from the train station after their intersection point, does not reduce the corrosion intensity, but may change the location where corrosion occurs on the pipeline. The following prevention measures are common:

1. The easiest measure is to increase the quality of pipeline coating, at least at regions where the pipeline goes parallel with the railway.
2. Another measure is to insulate the terrain surrounding the railway, preventing the stray currents from escaping to the nearby terrains.
3. Prevention of the rail resistance at connection points also reduces the potential of the surrounding terrain. Conductance of the rails is provided via the cable connections, and loose cables increase the potential, which leads to an increase in the stray currents escaping to the terrain.
4. Not allowing the stray current to come out of the pipeline, which tries to go back to the station, and thus normally leaves the pipeline at a location close to the station, would also prevent the corrosion. This can be done by placing an interference connection between the negative pole of the transformer unit and the pipeline. In case the pipeline

is not close to the transformer unit, then a metallic connection is placed between the railway and the pipeline, leading to stray currents go back to the transformer via the railway. The problem here is the determination of the resistance of these connections, since stray currents are only created when the train moves by, and the intensity of these currents vary as the train moves. Thus, the resistance must be automatically adjusted based on the varying current intensity; otherwise, at times when no train moves by, unnecessary current would be applied to the pipeline from the transformer unit.

5. If the pipeline is intersecting the railway, then the location where the current enters the pipeline can be definitely known; thus, entrance of stray currents can be prevented by applying impressed current cathodic protection to this region of the pipeline. Impressed current cathodic protection systems must be established at both sides of the intersection point, and must work only when there are trains moving by, which can be achieved via transformers that have automatic potential controls. In other words, when pipeline/terrain potential reaches a certain level due to the train that is passing by, the transformer unit begins operating, and stops as the train moves away. Pipeline/terrain potentials are continuously measured by a fixed reference electrode placed at the region where the interference effect of stray currents is maximum, which is also connected to the transformer unit, and thus signals it to begin operating when a certain when pipeline/terrain potential is reached.

### *ii. Interference Effect of High Voltage Transmission Lines*

A potential field is created due to the high voltage currents carried by the transmission lines leading to stray currents escaping to pipelines nearby from the earthed poles at the bases of the transmission lines. Intensity of the stray currents depends on the distance between the pipeline and the transmission line, the distance they go parallel to one another, and the coating quality of the pipeline. If the transmission lines were already present before the construction of the pipeline system, then it must be determined how much near can the pipeline system be constructed to the transmission lines and how long they can go parallel to one another.

Normally, the alternative current carried by the transmission lines that are escaping to the pipelines nearby do not cause corrosion; however, it is not the corrosion they cause here that is the problem, it is that the pipelines are not wanted to carry high potentials.

In the case that the transmission lines are being constructed after the construction of the pipelines, the following measures can be taken to prevent the harmful effects of escaping high voltage stray currents:

1. Surge protectors are placed at the beginning and ending points of the region of the pipeline that intersect or that go parallel with the transmission lines. Working potential of these surge protectors must be at least 2.5 kV.
2. A metallic shield or an old pipe can be placed at regions where the pipeline goes parallel with the transmission lines and closer to the pipeline.

### *iii. Insulated Flange Interference*

Insulated flange interference occurs when a potential difference is created at a terminal location of a pipeline such as at a point close to the tank, where pipeline is separated into two via an insulating flange. Interference corrosion occurs at areas close to the flange and at the part of the pipeline that is cathodically protected.

*iv. Ship-Pier Interference*

Interference effects are usually negligible in the sea, since seawater is very conductive. However, application of 20 A to cathodically protect a pier pole that is installed 100 m under the sea results in a 100 mV/1m current gradient, and when a ship approaches to such a pier that is cathodically protected, a potential difference up to 2 V to 3 V can be formed, exposing the ship to interference corrosion while the ship is at the pier. Further, the current needed for cathodic protection of the pier would also increase substantially.

*v. Stray Currents Formed Due To Welding Operations at a Ship*

When ships are welded using a welding machine that has its positive pole earthed at the pier, stray currents from the welding machine would go to the sea, and from thereof enter the body of the ship, resulting in interference corrosion at locations where they leave the ship. However, no corrosion would take place if the positive pole of the welding machine is earthed at the ship via an insulated cable, since then the stray currents originated from the welding machine that go to sea and enter the ship from thereof would come back to the welding machine via this cable.

## 8.10. Criteria for Cathodic Protection Projects

The following stages must be followed in an orderly fashion for successful installation and implementation of cathodic protection projects:

1. Size of the structure that will be cathodically protected is determined.
2. Electrochemical characteristics of the structure's surroundings are determined.
3. It is determined whether sacrificial anode cathodic protection or impressed current cathodic

protection would be more appropriate for the respective system.

4. Foreign metallic structures within the region of the structure to be protected are determined.
5. Regions along the pipeline are checked for anaerobic environments.
6. Service life of the cathodic protection unit is estimated.
7. Coating resistance is either theoretically estimated or experimentally determined.
8. Total current needed for cathodic protection is calculated.
9. Number of anodes and their locations are determined in the case of sacrificial anode cathodic protection systems.
10. In the case of impressed current cathodic protection systems, the type of the anodic bed, whether it will be a deep well or a shallow well anodic bed, is determined. If deep well anodic bed is preferred, then depth of the well, type and number of the anodes to be placed in the well are determined.
11. Anodic bed resistance is determined and service life of the anodic bed is estimated.
12. Types of connection cables and their cross sectional areas are determined.
13. Transformer/rectifier (T/R) unit's direct current potential is calculated.
14. Transformer/rectifier (T/R) unit's capacity and location is determined.
15. It is determined whether the transformer/rectifier (T/R) unit will be manual or automatic.

## 8.11. Cost of Cathodic Protection

Cost of cathodic protection systems consists of initial establishment costs, operational costs, and maintenance and repair costs. Initial establishment costs consist of the cost of the project and planning phase, costs of materials and equipment, installation costs, and the profit of the contractor. In impressed current cathodic protection systems, initial establishment costs are equal to operational costs of about 10 to 15 years, and thus are higher compared to sacrificial anode cathodic protection systems.

The major criterion determining the costs is the current needed for cathodic protection. A low cathodic protection current translates to low initial establishment and operational costs. However, the relation between these variables is not linear. Initial establishment costs per 1 Ampere current are higher for small T/R units with low current capacities than for large T/R units with higher current capacities. Among the items in initial establishment costs, the cost of materials and equipment is proportional with the current needed for cathodic protection, while the others are fixed costs. Cost of unit current intensity is very high for systems with current capacity of lower than 5 A, which decreases with increasing current capacity up to 20 A, while not much change in costs occur for units that have current capacities above 20 A. Costs per A.year also decrease with increasing cathodic protection service life.

It is also important to note that the alternative current received from the network that is converted to direct current to be used in cathodic protection systems does not have a fixed price as well. Cost of the alternative current received from the electricity network depends on the efficiency of the T/R unit, potential of the produced direct current, and cost of electricity power per kW.hour.

Costs can be decreased by reducing the current needed for cathodic protection. This can be done by increasing the coating quality, reducing the initial establishment and operation costs. A quality coating also results in a small attenuation coefficient, which translates to longer distances of pipeline that can be cathodically protected from one location, leading to a reduction in the number of T/R units needed, reducing the initial establishment costs per 1 km of pipeline. However, expenses for the coating must be balanced with the reduced costs of reduced current intensity, since the relation is not linear. Thus, usually a medium level coating quality would result in the most reduced costs.

In impressed current cathodic protection systems, anodic bed costs increase with increasing number of anodes. However, anodic bed resistance also decreases with increasing number of anodes, allowing the use of a lower capacity T/R unit to produce the cathodic protection current of the same intensity. Thus, there is an optimum for the number of anodes to be installed that accounts for the most economy, balancing out both the costs of the anodic beds and the T/R units.

## 8.12. Comparison of Cathodic Protection Systems

The major criteria to consider when choosing one of the two cathodic protection systems are current needed for cathodic protection and the resistivity of the terrain among other factors. Usually, if $i > 1$ Ampere and if $\rho < 3000$ ohm.cm, sacrificial anode cathodic protection systems are more appropriate. Other factors affecting the choice of cathodic protection system are:

1. *Presence of electricity power*: sacrificial anode cathodic protection is usually the single option in places where there is no electricity power, since it does not require an external current source. Galvanic anodes are the source of the required currents. If impressed current cathodic protection must be used, then power must be generated using a generator or via solar energy obtained through photovoltaic cells.
2. *Unit cost*: current produced by the sacrificial anode cathodic protection systems is more expensive than the current received from the network and converted to direct current by the T/R unit to be used for the impressed current cathodic protection systems. Thus, sacrificial anode cathodic protection systems are usually preferred if the current needed for cathodic protection is not high. Additionally,

initial establishment costs of impressed current cathodic protection systems are higher than the sacrificial anode cathodic protection systems; thus, if the planned duration of cathodic protection is short and the current needed for cathodic protection is low, sacrificial anode cathodic protection systems are preferred.

3. *Terrain/earth/ground resistivity*: circuit potential of sacrificial anode cathodic protection systems is usually low, and thus they are regularly applied in grounds with resistivity values up to 5000 ohm.cm, and not in fresh waters or in grounds with high resistivities unless a galvanic anode with a high potential is used. On the other hand, impressed current cathodic protection systems can be employed in grounds with high resistivities, which can be achieved either by reducing the anodic bed resistance using deep well anodic beds in dry soils or by increasing the cathodic protection current.
4. *Ease of applicability*: application of sacrificial anode cathodic protection systems is easier than application of impressed current cathodic protection systems. If current need increases due to unpredictable factors in time exceeding the capacities of the existing anodes, then more anodes can be installed to produce higher total current intensities, while in impressed current cathodic protection systems T/R units cannot be adjusted to work over capacity and anodic bed resistance cannot be reduced during operation; thus, such measures should be taken in the beginning during the project phase.
5. *Cathodic protection current adjustment*: if the current need for cathodic protection increases in sacrificial anode cathodic protection systems, the potential of the system to be protected decreases, leading to a higher margin between the potentials of the anode and the cathode, resulting in a higher current

withdrawal from the anode, while in impressed current cathodic protection systems, either the T/R unit has to be adjusted manually or a reference electrode may be installed at the cathodic protection system connected to the T/R unit, leading to the T/R unit automatically adjust the current. If the appropriate current adjustments are not made in a timely manner, cathodic protection may not be provided, or overprotection may occur.

6. *Dissolution of the cathodically protected metallic structure*: in sacrificial anode cathodic protection systems, dissolution is not observed in the pipelines, in areas that are close to the anode due to presence of high potentials, while dissolution is observed in impressed current cathodic protection systems.
7. *Interference effects*: interference effects are negligible on the nearby metallic structures in galvanic cathodic protection systems, since anode-ground potential is low. On the other hand, stray currents may originate from the impressed current cathodic protection systems, leading to interference effects; thus, surrounding structures must be periodically monitored and corrosion preventative measures must be taken.
8. *Maintenance*: periodical controls of impressed current cathodic protection systems are easier since T/R units are usually at easily accessible locations and the current can be adjusted from one location; thus, the overall system can be controlled from one location, while in the case of sacrificial anode cathodic protection systems, each anode must be checked individually.
9. *Replacements*: in sacrificial anode cathodic protection systems, when there is a problem, the anode that has the problem is found and replaced, while in impressed current cathodic protection systems, when there is a problem at the anodic bed, whole system may need to be replaced.

10. *Electrical connections*: cable connections in sacrificial anode cathodic protection systems are like any other, while cables connecting T/R unit to the anodes in impressed current cathodic protection systems must be insulated very well, since even a small gap may result in detachment of the connection to the anode. Besides, there is no risk of wrong connection in galvanic anodes, while T/R unit's negative pole must be connected to the metallic structure to be protected and its positive pole must be connected to the anodic bed in impressed current cathodic protection systems; if it is not, then the metallic structure to be protected corrodes away shortly.

Despite its advantages, cost per unit current is high in sacrificial anode cathodic protection systems, and thus it is not preferred in cases where high currents are needed. Since galvanic anodes have low potentials, they cannot be used in electrolytes with high resistivities, and if for any reason during operation, the cathodic protection current intensity rises to a value exceeding the current capacities of the galvanic anodes, new anodes must be installed.

Besides, anodes make up half of the initial establishment cost of cathodic protection systems, and thus feasible choice of the anode is important. Current withdrawn from the unit area of the anode must be as high as possible and anodic resistance must not increase in time; also, weight loss per A.year current withdrawn from the anode must be as small as possible. Additionally, anodes must be durable, light, and inert to environmental chemical factors. One also must be careful when administering impressed current cathodic protection, since substantial rises in potentials may lead to hydrogen evolution, leading to hydrogen embrittlement and coating failures.

# 9

# Sacrificial Anode or Galvanic Cathodic Protection Systems

Cathodic protection changes the metal's potential in cathodic direction, preventing its dissolution. There are two methods of changing metal's potential in negative direction, providing an external current with one of them being sacrificial anode, or galvanic cathodic protection. In this method, an auxiliary anode that is more electronegative than the metal to be protected is connected with the metal. As electrons flow from the anode to the metal, the metal ions do not leave the metal, and corrosion is prevented. Anode sacrifices itself for protecting the metal, as it is used for protecting underground structures as well as ships. To exemplify, galvanic anodes produce current similar to a battery, which is carried to the cathode through an external connection, providing the electrons needed for the cathodic reaction. If galvanic cathodic protection cannot be efficient due to the characteristics of the environment, impressed current cathodic protection should be preferred. Galvanic anodes must be negative just enough to cathodically polarize the metal to be protected.

## 9.1 Anodic Potentials and Anodic Polarization

Galvanic anode's potential must be negative enough to cathodically polarize the metal to be protected, since it is the potential difference between the anode and the cathode that allows the cathodic protection current to flow. This potential difference must overcome the resistance of the cathodic protection system. Anodes with low potentials cannot be used in terrains or grounds with high resistivities. In the case of iron and steel structures, cathodic potential must be decreased to below –850 mV compared to copper/copper sulfate (CSE) reference electrode. Additionally, the IR ohmic potential reduction that is primarily due to the anodic bed resistance, resistance of the connection cables, and the ground resistivity must be taken into consideration when measuring the cathodic potential under the applied current and calculating the total current needed for cathodic protection.

Galvanic anodes must dissolve uniformly for a uniform production of the cathodic protection current, and while this is possible in the seawater, anodes polarize promptly underground. As a reason, metal ions formed via dissolution of the anodes precipitate mainly as hydroxides on the anode's surface, reducing the active surface area and increasing the anodic resistance, leading to an increase in potential in the positive direction; hence, produced current decreases in time. Therefore, appropriate anodic beds must be used to prevent such polarization to take place.

## 9.2 Galvanic Cathodic Protection Current Need

The required current to be provided by the anode depends on factors such as the total surface area of the metallic structure to be protected, the quality of the surface coating of present, anodic open circuit potential with respect to the metallic structure to be protected, and the anodic bed's or electrolyte's resistivity or impedance, as well as anode's resistivity in the electrolyte or the anodic bed.

If the service life of the anodes is longer than the planned duration of cathodic protection service life, then smaller anodes, or in other words, anodes with higher length vs. diameter (L/d) ratios, can be used, or a different type of anode such as HP magnesium anodes can be used instead of AZ-63 magnesium anodes, since they can produce the same amount of current with less number of anodes for shorter durations.

On the contrary, if the service life of the anodes is shorter than the planned duration of cathodic protection service life, then bigger anodes, or in other words, anodes with smaller length vs. diameter (L/d) ratios, can be used, increasing the anodic mass but without increasing the produced current intensity. Another way is installation of multiple anodes parallelly that are connected from one location to the pipeline, thus increasing the anodic bed resistance and decreasing the unit current withdrawn from an anode.

Normally, the current withdrawn from the galvanic anodes is decreased in time due to formation of protective layers of corrosion products, water hardness, etc. on the surface of the metallic structure to be protected, leading to longer than theoretically predicted service lives. However, it must be taken into consideration that in some cases, higher cathodic protection current needs necessitated by the damaged coating, which wears out in time, may balance out the former effect.

## 9.3 Anodic Current Capacity and Anodic Current Efficiency

Anode current capacity is the current that 1 kg of the anode can produce in one hour (A.hour/kg), which varies with the amount of current withdrawn from the anode ($mA/m^2$) and temperature:

$$\text{anodic current capacity}\,(A.hour/kg) = \frac{\text{anodic current production}\,(A.hour)}{\text{anodic mass}\,(kg)} \quad (51)$$

In practice, another version of the same term that is the amount of the anode, which can produce 1 Ampere per year (kg/A. year), is mostly used.

The aforementioned theoretical current capacities are calculated based on the Faraday Law, while in practice, the current capacities of the anodes are smaller. The ratio of the actual or true current capacity to that of theoretical current capacity is known as anodic current efficiency:

$$\text{anodic current efficiency} = \frac{\text{true current capacity}}{\text{theoretical current capacity}} \times 100 \quad (52)$$

Anodic current efficiency depends on the type of the anode and the current density withdrawn from the anode. Magnesium anodes have 50% to 60% anodic current efficiencies, while zinc and aluminum anodes have up to 90% efficiencies.

## 9.4 Service Life of an Anode

In practice, only a certain percentage of an anode's mass can be used for current production, since as the anodes are used, their surface area gets smaller, increasing the anodic resistance, and also due to the non-uniform consumption of anodic mass; thus, anodes can never be fully consumed. This percentage of anodic mass that can be used is called the usage factor. Usage factor primarily depends on the anode's shape, which is 85%

for cylindrical galvanic anodes, for instance. Based on the usage factor, anodic current capacity can be calculated via formula 50 and anodic current efficiency can be calculated via formula 51; thus, the life an anode can be estimated, given that the cathodic protection current need and the mass of the anode are known:

$$\text{life of an anode (year)} = \frac{\text{anodic mass}(kg) \times \text{usage factor} \times \text{anodic current efficiency}}{\text{anodic current}(A) \times \text{theoretical current capacity}(kg/A.year)} \quad (53)$$

## 9.5 Minimum Number of Galvanic Anodes

The number of anodes that will be installed must satisfy both the total current needed for cathodic protection in terms of Amperes and also the total anodic mass. In the first method, number of anodes (n) is determined, dividing the total current needed for cathodic protection (I) with the anodic current intensity or the current output of each anode (i):

$$n = I/i \quad (54)$$

Alternatively, number of anodes (n) is determined by dividing the total required anodic mass (M) with the mass of a single anode (m):

$$n = M/m \quad (54)$$

and the total anodic mass is determined from the anodic life formula, formula 52, since all variables are known but the anodic mass. Consequently, out of the two calculations, the bigger number of anodes is chosen and installed.

## 9.6 Commonly Used Galvanic Anodes

Galvanic anodes are chosen among metals or alloys that are more active than the metal to be protected and are directly connected to the protected metal, forming a combined potential that is negative compared to the system's potential. This way, system's potential becomes $E'_{corr}$ instead of $E_{corr}$, and $i'_{corr}$ instead of $i_{corr}$. Since in this method, the galvanic anode corrodes instead of the metal to be protected, it has a limited service life.

Galvanic anodes have certain current efficiencies and current capacities. By installing sufficient quantity and number of anodes in the cathodic protection system, a metallic structure can be kept as the cathode for the desired amount of time, and thus protected from corrosion. Anode quantity and size can only be determined based on the current needed for cathodic protection. For galvanic anodes to provide protection even in high resistivity environments, equilibrium potential must be very negative. The anode should not be polarized too much, and only a little electrochemical mass loss should result in high currents, indicating high efficiency.

Consequently, galvanic anodes' corrosion potentials must be sufficiently negative, their anodic capacities and anodic efficiencies must be high, they should be continuously active and not passivated, etc. If more than one anode is connected to a pipeline at the same location, they must be connected parallel to one another via one main cable. It is more expensive to dig holes for each anode and connect those using different cables, because it requires more labor, and also there may not be sufficient number of suitable locations along a pipeline to separately install the anodes.

Zn, Al, and Mg alloys are the most commonly used sacrificial anodes. The percentage of noble metals present in these alloys cannot be more than a certain limit, since they would passivate them, making them useless as sacrificial anodes. Al has the highest current efficiency of 95%, leading to a theoretical current capacity of 2830 A.hour/kg, Mg has a low current efficiency of 55%, leading to a relatively high theoretical current capacity of 1230 A.hour/kg, and Zn has a high current efficiency of 95%, leading to a low theoretical current capacity of 780 A.hour/kg. Additionally, only about 85%–90% of the theoretical current capacities can be utilized, which is called the "usage factor."

The anode is chosen based on the environment's resistivity; e.g., Zn or Mg is used in highly resistant environments, while

Al is used in environments that have low resistivities. Thus, zinc and magnesium alloyed anodes are preferred for cathodic protection of the metallic structures underground such as pipeline systems, while aluminum anodes are preferred for cathodic protection of the metallic structures in the seawater. Magnesium anodes and especially high potential (HP) magnesium anodes are preferable in terrains with high resistivities and in fresh waters. Zinc anodes are inexpensive and have high anodic current efficiencies; however, they can only be used in terrains with resistivities lower than 2000 ohm.cm and in salt waters, since their circuit potentials are low. Magnesium anodes are not suitable for terrains with resistivities higher than 5000 ohm.cm, while magnesium anodes can directly be placed in the ground in terrains with resistivities lower than 500 ohm.cm. Although magnesium anodes' anodic current efficiencies are low, their anodic current capacities are higher those of zinc anodes' under same conditions. On the other hand, aluminum anodes have very high current capacities, and thus they are the least expensive, e.g., 3.5 kg aluminum would be enough for producing 1 A.year current, while the same 1 A.year can be produced with 7.88 kg magnesium and 11.84 kg zinc. However, since aluminum anodes passivate even when they are alloyed with mercury and indium, they can only be used in seawater.

The costs of anodes are compared in terms of the costs of unit masses along with the current capacities, which account for the cost of production of 1 Ampere for one year. For instance, the approximate unit price of magnesium, considering late 2012 prices, is $3.20/kg, and considering magnesium anodes' true current capacity of 1100 A.hour/kg, the cost of 1A.year is calculated as around $23.

Current withdrawn from the galvanic anodes can be increased depending on the potential difference between the anode and the cathode and also on the anodic bed resistivity, which can be achieved by increasing the surface area using smaller anodes, such as two small anodes compared to one big anode that has the mass of the two small anodes; thus, it is more appropriate to use many small anodes instead of a few large anodes, especially in grounds with high resistivities.

Another issue is the possibility of overprotection. Zinc anodes have low circuit potentials: the system's potential can only be increased up to –1.0 V using zinc anodes, and when that potential is reached, the withdrawn current is automatically reduced, and thus the problem of overprotection is never encountered. In the case of magnesium anodes, however, produced potentials may be higher than needed, resulting in excessive currents, resulting in overprotection and the anode to be wasted. Consequently, since zinc anodes' service life is longer than the service life of the magnesium anodes' of same size, they are preferable when the expected cathodic protection duration is long, and vice versa.

### 9.6.1 Magnesium Anodes

Magnesium anodes' electrode potential is relatively less compared to metals that are at the top of electrode potential series due to magnesium's thin protective oxide film on its surface. Additionally, magnesium anodes form $Mg(OH)_2$ in aqueous solutions that have no chlorides or sulfates present. $Mg(OH)_2$ has very low solubility; it precipitates on the surface of the anode and passivates it. In presence of chlorides and sulfates, $Mg(OH)_2$ film is damaged and the anode dissolves in the form of pitting corrosion. Thus, magnesium anodes cannot be used in their pure form. Pure magnesium's electrode potential is –2.4 V compared to standard hydrogen electrode, while in seawater its potential reads –1.55 V compared to copper sulfate (CSE) reference electrode. By adding some manganese of at least 0.5%, potential of magnesium anodes can be increased up to –1.75 V, since manganese cancels iron's harmful effects. Therefore, these high potential magnesium anodes can be used in high resistivity terrains and in fresh waters. Addition of aluminum and zinc also positively affect magnesium anodes' potentials and current capacities, while impurities such as iron, nickel, copper, and silicon have negative effects on magnesium anodes' potentials and current capacities. Withdrawal of high currents during cathodic protection increases the anodic current efficiencies of magnesium anodes.

Magnesium anodes are usually produced in the shape of a bar or a "D" letter, and they have a steel skeleton along their axis. They are sold either as bare anodes or as packaged with anodic bed materials. High potential (HP) magnesium alloy is among the most commonly used magnesium anodes. Percentage of the manganese in the HP alloy depends on the percentage of the aluminum, e.g., Galvomag high potential (HP) magnesium anode has 1.25% manganese and 0.1% aluminum. Thus, HP alloys in general have around 0.05% aluminum and manganese in the amount of at least [0.5% + (0.6 × aluminum percentage)]; thus for 0.5% aluminum, manganese needs to be 0.8%, while 0.03% zinc, 0.05% silicon, 0.02% copper, 0.03% iron, 0.02% nickel, and magnesium make up the remaining amount. HP alloy's circuit potential compared to steel is 900 mV.

Another magnesium alloy used as a galvanic anode is AZ-63 alloy, which has a density of 1.7 $g/cm^3$ and a 700 mV circuit potential compared to steel. Magnesium AZ-63 alloy is obtained via addition of 5.3% to 6.7% aluminum, 2.5% to 3.5% zinc, 0.25% to 0.40% manganese, 0.3% silicon, 0.08% copper, 0.03% iron, 0.03% nickel, and magnesium making up the remaining amount. Adding sufficient amount of manganese to the alloys results in high potential AZ-63 alloy anodes, which are suitable for terrains with high resistivities. Theoretical current capacity of AZ-63 magnesium alloy anode is 2200 A.hour/kg or 3.94 kg/A.year, which is higher than other anodes; however, due to low anodic current efficiency of 50%, true current capacity becomes 1100 A.hour/kg or 7.88 kg/A.year. AZ-63 alloy has a better current efficiency than the HP alloy. If the withdrawn current density is less than 0.3 $mA/cm^2$, then HP magnesium anodes' current efficiency decreases even below 50%, while the current efficiency for AZ-63 magnesium alloy anodes are more than 60% at the same current density.

### 9.6.2 Zinc Anodes

Zinc anodes were first used in 1824 by Sir Henry Davy for the purpose of cathodic protection of ship bodies that are made of copper. Zinc anodes are still commonly used for the cathodic protection of marine structures and structures in terrains with low resistivities. Zinc anodes have at most 0.006% lead, 0.005% iron, 0.005% copper, 0.15% cadmium, 0.50% aluminum, and 0.125% silicon as the alloying elements, with zinc making up the remaining amount.

Pure zinc anode's potential in seawater is around –1.10 V compared to copper sulfate (CSE) reference electrode, resulting in a driving voltage of around 250 mV, compared to commonly accepted protection potential of steel in an aerobic electrolyte of nearly neutral pH, which is –850 mV, which is not sufficient for cathodic protection in fresh waters and underground with resistivities higher than 2000 ohm.cm. However, this potential difference decreases if the zinc anode contains some iron impurity, and thus the amount of iron in zinc anodes should be less than 0.0014% if no aluminum is present, and less than 0.0030% if there is about 0.1% aluminum present in the alloy, which cancels the negative effect of iron. An example is MIL-A 18001 anode that has aluminum in the amount of 0.1% up to 0.3%, which allows formation of alloys with iron up to 0.003%, and thus it can be used in seawater. In any case, it is important for the iron percentage to be under 0.005%, since higher percentages of iron lead to the formation of iron oxide layer, passivating the surface of the anode. Aluminum binds the iron in the anode, preventing it from becoming the cathode of a corrosion cell within the anode of which zinc is the anode. In the absence of aluminum, zinc forms primarily zinc hydroxide and other corrosion products, which cover the surface of the anode, leading to its passivation, and thus reducing its potential.

Impurities such as lead and copper negatively affect zinc anodes' efficiency as iron does, while aluminum and cadmium have positive effects. Cadmium cancels lead's poisoning effect as aluminum cancels iron's. MIL-A 18001 anodes have 0.06% cadmium, up to 0.006% lead, and 0.005%. copper.

Zinc anodes used underground do not contain aluminum and also have less cadmium and iron, as opposed to zinc anodes used in seawater. Zinc anodes used underground dissolve in time, producing zinc hydroxide, which precipitates on the surface of the anode, leading to passivation, especially if ions such as carbonates, phosphates, and silicates are present in the medium, which is not the case in the seawater, since the produced corrosion products of zinc dissolve into the seawater and do not precipitate on the surface. Thus, zinc anodes are placed in anodic beds to be able to be used underground. Anodic bed material contains sodium and calcium sulfates, and since zinc sulfate dissolve easily as zinc chloride does, anodes cannot passivate. Further, zinc hydroxide forms a gel on the surface of the anode at high temperatures, especially over 60°C.

Zinc anodes have around 820 A.hour/kg theoretical current capacities and high anodic current efficiencies of over 90%, leading to 738 A.hour/kg or 11.84 kg/A.year true current capacity. Their density is 7.1 g/cm$^3$, and the cost of 1A.year current production comes to around \$14; thus, it costs about 60% less than magnesium. Current capacity of zinc anodes increases with the withdrawn current; however, it decreases with increasing temperatures.

### 9.6.3 Aluminum Anodes

Aluminum anodes are used in seawater and in other waters that are slightly salted. Presence of copper and nickel in aluminum anodes shift the potential of aluminum in the positive direction, while the presence of zinc, magnesium, and cadmium decrease the passivation and shift the potential of aluminum in the positive direction. Additionally, the presence of mercury, tin, and indium keep aluminum anodes active at all times, leading to uniform dissolution of the aluminum anodes.

Normally, aluminum is more active than zinc when electromotor force series is considered; however, due to the naturally occurring protective oxide film on its surface, it could not be used as an anode until the 1950s, when it had a potential of –900 mV compared to copper sulfate (CSE) reference electrode in seawater. After the 1960s, the addition of about 3% zinc and

0.5% tin improved current efficiency of aluminum anodes in seawater up to 50% and potential up to –1.3V, while alloying elements such as Zn and Hg or Zn and In increased the current efficiency up to 90% and the potential up to –1.05V.

Alloying elements of Hg in the amount of 0.03% to 0.05%, In in the amount of 0.01% to 0.03%, and trace amounts of Sn prevents passivation of aluminum. Indium alloyed anodes are preferred over mercury and tin alloyed anodes in seawater, since both are toxic and cause environmental pollution. Indium alloyed aluminum anodes have 2.1–2.7% zinc, 0.003% to 0.032% cadmium, 0.01% to 0.15% silicon, at most 0.25% iron, and 0.017% to 0.024% indium, with aluminum making up the remaining amount. Indium is nobler than aluminum, and thus, it keeps aluminum active via the following reaction:

$$Al + In^{3+} \longrightarrow Al^{3+} + In \qquad \text{(Eq. 89)}$$

Mercury alloyed aluminum anodes have 0.35% to 0.50% zinc, 0.11% to 0.21% silicon, at most 0.25% iron, and 0.035% to 0.45% mercury, with aluminum making up the remaining amount. Since iron passivates aluminum anodes, its amount must not exceed 0.1% in the alloy. Silicon cancels the negative effects of iron, and thus a Galvalum alloy that has iron up to 0.15% could be used as anodes with the addition of silicon.

Despite having a relatively low potential of –1.10 V in seawater that has 25 ohm.cm resistivity at 25°C for 300 mA/$m^2$ current density withdrawn leading to 250 mV circuit potential compared to steel, aluminum anodes have 2.4 times more current capacities than magnesium anodes and 3.6 times more current capacities than zinc anodes, equaling to 2960 A.hour/kg theoretical current capacity and 2670 A.hour/kg or 3.5 kg/A.year true current capacity, due to up to 90% current efficiencies. The density of aluminum anodes is 2.7 g/$cm^3$, and they are about three times less expensive in terms of produced unit current compared to zinc and magnesium anodes; however, aluminum anodes can only be used in seawater or in salt waters with low resistivities, since resistivities higher than 50 ohm.cm for

Al + Zn + Hg alloys and higher than 500 ohm.cm for Al + Zn + In +Si alloys cause the potential to go below –1.0 V.

## 9.7 Performance Measurements of Galvanic Anodes

Galvanic anodes' electrochemical properties are estimated before their installations to establish whether they can produce the desired current capacities, current efficiencies, and potentials for the planned duration of cathodic protection. These chemical properties are as follows:

### 9.7.1 Chemical Composition

Anodic material must be checked for alloying elements or impurities that negatively affect the anode's performance.

### 9.7.2 Mechanical Strength

Mechanically, the weakest parts of the anodes are the cable connections, which must endure a weight that is five times more than the anode's mass and a maximum weight of 100 kg. Under such weight, the cable must not be damaged or broken.

### 9.7.3 Electrical Resistivity

A steel skeleton is present along the inner axis of the galvanic anodes for conductance. This skeleton and the anode must be entirely fused together not to yield any electrical resistance inside the anode, which would otherwise lead to a reduction in anodic current production. This electrical resistance inside the anode is determined via applying a current of 1 A from an external direct current source and measuring the occurring potential differences in intervals of 0.1 mV. The same measurement is repeated with currents of 2, 3, 4, and 5 Amperes. Electrical resistance inside the anode should not be more than 0.01 ohm.

### 9.7.4 Electrochemical Experiments

Potentials of the anodes are usually measured in clean seawater or in seawater with regards to ASTM 1141 standards at 20°C, compared to silver/silver chloride reference electrode (SCE) or copper/copper sulfate electrode (CSE). If measured against CSE electrode, then –70 mV is added to the potential measured with that of SCE. Current capacity and current efficiency measurements are either done by galvanostatic method or by free current output method.

*i. Galvanostatic Method*

Galvanostatic method is based on measurement of the weight loss, which is due to withdrawal of current from the anode. For this purpose, anode surfaces are cleaned with concentrated nitric acid, removing the oxide layers, and washed afterwards with distilled water. A constant current density that is between 0.5 and 0.7 mA/cm$^2$ is passed through the anodes for 5 to 10 days, until the past current reaches 0.1 A.hour. Then, the anode is washed free of corrosion products and weighed for weight loss determination. The following formula is used to determine the true current capacity;

$$true\ current\ capacity\left(A.\frac{hour}{kg}\right) = \frac{current\ passed\ through\ the\ circuit(A.hour)}{anodic\ mass\ loss(kg)} \quad (56)$$

and then the true current capacity value is inserted into the formula 51 to determine the anodic current efficiency. For instance, for a mass loss of 40 mg with 0.10 A.hour passed current, true current capacity is measured as:

$$true\ current\ capacity(A.hour/kg) = \frac{0.10(A.hour)}{40 \times 10^{-6}(kg)} = 2500\ \text{A.hour/kg}$$

and considering the theoretical current capacity of aluminum anodes, which is equal to 2965 A.hour/kg, anodic current efficiency is found as:

$$\text{anodic current efficiency}(\%) = \frac{2500(A.hour/kg)}{2965(A.hour/kg)} \times 100 = 84.3\%$$

*ii. Free Current Output Method*

For the free current output method, first, a cathodic protection circuit that has a cathodic area/anodic area ratio of at least 200 is established. Mild steel is used as the cathode, and synthetic seawater is used as the electrolyte. Current passes through the circuit due to the driving voltage between the anode and the cathode, and thus application of an external current is not required. Potentials and currents are measured and graphed vs. time for two weeks, and weight loss is calculated. Current withdrawn during the experiment is calculated via taking the integral of the area below the current intensity vs. time curve using the following formula, where i is the current, t is time, and Q is charge in Coulombs:

$$Q = i \times t \tag{57}$$

## 9.8 Galvanic Anodic Beds

Galvanic anodes are not directly buried in ground, but placed in anode beds, so that the anode is uniformly dissolved, leading to a high usage factor. Furthermore, the anode's surroundings constantly remain wet, leading to a decrease in anode resistance, resulting in a higher current output. The conductive salts anodic beds contain also reduce the anodic resistance, leading to low electrical resistance, preventing anodic polarization. Consequently, anodic beds make it possible for the anodes to be used in terrains with high resistivities.

In areas with high resistance, anode beds are specially prepared. Anodes should be casted and their shape must be appropriate to the protected structure. Their connections with the system must be such that electrical conductivity is good and mechanical durability is high.

Usually, galvanic anodes are commercially produced and available in packages; however, if not, first, one-third of the anodic bed filling material is placed in the anodic bed, then the anode is placed in the middle of the dug hole, and the remaining anodic bed filling material is used to surround the anode filling the anodic bed. Prepackaged magnesium anodes are commonly placed 1.5 to 3 meters away and 1 meter below the pipeline or until the wet region is reached. This way, anodes are not affected by the meteorological events and foreign construction operations in the nearby areas. Galvanic anodes are connected to pipelines in a sequential manner in equal distances. Placing the anodes into locations where the ground resistivity is the lowest is more appropriate. If more than one anode will be connected to the pipeline at the same location, first, anodes are connected parallel to each another, and then they are welded to the pipeline as a group using thermite welding. The welding location and other connections must be insulated well, and cable cross sections must be at least 6 $mm^2$, so that no decrease in potential occurs.

### 9.8.1 Anodic Bed Filling Materials

There are usually two types of anodic bed filling materials: type-A consists 70% to 75% gypsum ($CaSO_4.2H_2O$), 20% to 25% bentonite, and 5% to 6% sodium sulfate, leading to a unit electrical resistance of 50–100 ohm.cm, while type-B consists 40% to 50% bentonite, 25% to 30% gypsum ($CaSO_4.2H_2O$), and 25% to 30% sodium sulfate, leading to a unit electrical resistance of 25–50 ohm.cm. Type-A filling materials are more appropriate for magnesium, while type-B filling materials are more appropriate for zinc anodes. Solubility of gypsum present in the anodic bed materials is 3 g per liter of water, and it continuously releases sulfate ions to the substrate surface, preventing the formation of hydroxide film for long periods of time, maintaining low resistivity. Sodium sulfate's solubility is very high, and its use reduces the resistivity to below 100 ohm.cm. Besides, bentonite can absorb high amounts of water, keeping the anodic bed wet.

### 9.8.2 Anodic Bed Resistance

One of the most important parameters in the design of cathodic protection systems is the electrical resistivity of the environment. Resistivities encountered for pipeline environments, for instance, vary from 1 ohm cm in brackish river water to more than 500,000 ohm cm in non-porous granite. Measurement of the resistivity of the environment and calculation of the electrical resistance between the anodes and the structure due to the electrolyte must be made at an early stage in the design of the scheme to ensure that adequate current output will be obtained from the anodes over the service life of the structure. However, exact calculation of the electrical resistances between anodes and structure due to the electrolyte is rarely possible.

Anodic beds can be considered to lie in a semi-infinite electrolyte, and the resistances of electrodes to infinite earth or seawater can be calculated for a number of anode shapes. Thus, if the anodes are remote from the structure, these anode resistances can be used to determine current output of anodes using Ohm's Law along with the difference between the required protection potential and the anode potential:

$$i = V/R \tag{58}$$

where i is the current output of the galvanic anode in Amperes, V is the measured potential difference in Volts, and R is the resistance of the anode or anodes connected parallelly. If, however, the anodes are positioned close to the structure, then some correction to the resistance is required. When current flows from a small anode to a large metallic structure, the current density is at a maximum near the surface of the anode. Hence, a major portion of the potential drop between anode and structure occurs in the vicinity of the anode, which allows values of anode resistance to infinite earth to be used reasonably successfully even when anode and structure are not well separated.

In practice, the distribution of current to a metallic structure that is to be cathodically protected is difficult to control. When the protection of a pipeline system by an anode is considered, for instance, it should be clear that there will be a higher current density at the point on the pipeline nearest the anode than elsewhere. Thus, it is obvious that the center of the pipe nearest the anode will be overprotected to some degree to ensure that the ends of the pipe are protected as well. This effect can be minimized by using several anodes spaced along the pipe, which, however, would greatly increase installation costs.

Theoretically, anodic bed resistance depends on the size of the anode, its diameter and length, and on the resistivity of the terrain. It consists of the resistance of an anode from the metal to the anodic bed, and the resistance of the anodic bed that is from the anodic bed to the terrain or to earth. Both resistances are first calculated separately using the Dwight formulas, and then are added together to find the total anodic bed resistance. Resistance of the anode or anodes depends on the way the anodes are placed in the dug holes, either vertically or horizontally, and also on the number of the anodes that are installed and the way they are connected to one another, while the resistance from the anodic bed to the terrain or earth depends primarily on the anodic bed filling materials.

### *i. Resistance of a Single Anode*

Resistances of anodes differ based on their vertical or horizontal placement. Resistances of the anodes that are placed vertically are higher than those that are placed horizontally. Resistance of a single cylindrical anode that has a length vs. diameter ratio of higher than 5 ($L/d > 5$) and is entirely embedded in the ground, either horizontally or vertically, is calculated via the following H. B. Dwight formulas:

$$R_{\text{vertical}} = \frac{\rho}{2\pi L}\left(\ln\frac{8L}{d} - 1\right) \tag{59}$$

$$R_{\text{horizontal}} = \frac{\rho}{2\pi L}\left(\ln\frac{4L}{d} - 1\right) \tag{60}$$

where single anode resistance (R) is in ohm, anodic bed resistance ($\rho$) is in ohm.cm, and both anode length (L) and anode diameter (d) are in cm. For bar anodes that do not have circular cross section, effective diameter must be used in the place of the diameter in the formula 55, which can be calculated using the following formula:

$$\text{effective diameter (d)} = \sqrt{\frac{cross\ sectional\ area}{\pi}} = \sqrt{(r_1 x r_2)/\pi} \quad (61)$$

*ii. Resistance of Multiple Anodes*

Resistance of a group of anodes that are connected parallel to one another and then as a group to the pipeline is calculated via the following formula:

$$R = \frac{\rho}{2\pi L n}\left(\ln\frac{8L}{d} - 1 + \frac{2L}{s}\ln 0.656n\right) \quad (62)$$

where n is number of the anodes and s is the distance between the anodes in cm.

## 9.9 Sacrificial Anode Cathodic Protection Projects

There are basic steps to be followed when realizing a project of a sacrificial anode cathodic protection system:

1. First, the surface area of the metallic structure to be cathodically protected is calculated.
2. Then, the amount of cathodic protection current need per area is determined based on field experiments.
3. Total current needed for cathodic protection is obtained by multiplying the cathodic protection current needed per area measured in step 2 and the actual surface area measured in step 1.
4. Afterwards, it is decided whether it is more appropriate to use sacrificial anode cathodic protection or impressed current cathodic protection, based on the resistivity of the terrain and on the total current

needed for cathodic protection, which is calculated in step 3.

5. If sacrificial anode cathodic protection is more suitable, then type of the galvanic anode is chosen based on the resistivity of the terrain.
6. Resistance of the anodic bed is calculated via sum of the resistance of the anode or anodes and the anodic bed backfill material based on formulas 58 through 61.
7. Circuit potential of the anode compared to the metallic structure to be protected is divided with the anodic bed resistance, resulting in the anodic current intensity that can be withdrawn from an anode based on formula 57.
8. Then, the anodic current output found in step 7 is inserted into formula 52 to find the life of an anode, or in other words, duration of cathodic protection, given that a certain type of galvanic anode is used.
9. Number of anodes is calculated via two methods via formulas 53 and 54 and of the two numbers; the bigger number is chosen, and that many anodes are installed.

## 9.10 Maintenance of Sacrificial Anode Cathodic Protection Systems

Sacrificial anode cathodic protection systems should be checked once a month for the first year the system is fully operational; afterwards they can be checked twice a year, of which one check should be performed during the rainy season. For the purpose of the maintenance, the following measurements and controls should be performed:

- Measurement of pipeline/terrain potentials at both "on" and "off" positions,
- Measurement of anode/terrain potential at "off" position,

- Measurement of system potential or the driving potential,
- Measurement of current intensity withdrawn from the anodes,
- Control of measurement units and connections,
- Control of insulations of insulated flanges, and
- Control of resistances of stray current connections if present.

The following problems are the most-encountered problems in sacrificial anode cathodic protection systems that need to be checked for:

### 9.10.1 Low Pipeline/Terrain Potential

Pipeline/terrain potentials that are not more negative than –850 mV, despite high anodic current withdrawal, may be due to one or more of the following reasons:

1. A new component may be added to the pipeline system; however, it may not have been separated using an insulated flange, and thus increased the current need.
2. Insulated flanges may have lost their resistances, resulting in stray currents escaping to foreign structures.
3. A new cathodic protection system that has a more negative potential may be established in the surroundings, resulting in stray currents escaping to this new system.
4. Pipeline coating is damaged to the extent that the increase in the current needed for cathodic protection is at a level that was not initially predicted and planned for.
5. Corrosive characteristics of the terrain, where the pipeline is located, have changed and become more corrosive due to increasing humidity levels and/or increasing diffusion rate of oxygen.

### 9.10.2 Decreases in Anodic Current Production

Normally, decreases in anodic current production occur gradually due to the current being withdrawn from the anode, leading to losses in weight and volume, and also due to the decreasing driving voltage caused by the increase in pipeline/terrain potential. However, in both cases, as long as the pipeline/terrain potential remains above the protection limit, cathodic protection continues; but if no anodic current can be produced and pipeline/terrain potential remains below the protection limit, then one or more of the following circumstances may be the reason:

1. More than predicted current is withdrawn from the anode, and thus the anode is detached before its estimated service life.
2. Anode-cable connection is weakened or detached completely; if detached completely, anode/terrain potential cannot be measured.

# 10

# Impressed Current Cathodic Protection Systems

It is possible to convert the metallic structure into a cathode using a direct current of right intensity and potential via impressed current cathodic protection protecting it from corrosion. Direct current used for cathodic protection is obtained through a transformer/rectifier unit, where alternative current is converted to direct current. The plus pole of the direct current coming out of the T/R unit is connected to the anodic bed and the negative pole is connected to the metallic structure, e.g., to the pipeline to be protected. Intensity of the current and the system's potential can be increased as much as desired in impressed current cathodic protection systems up to the capacity of the T/R unit, as opposed to the sacrificial anode cathodic protection systems, which require installment of additional anodes in case total cathodic protection current need increases. Thus, very long pipelines can be protected using just one anodic bed in impressed current cathodic protection systems, provided that pipeline/terrain potential does not exceed a certain limit value at the point of application, so as not to cause overprotection that cannot be tolerated. The length of a portion of a pipeline that can be protected using a single anodic bed depends upon the type of the pipeline, its inner and outer diameters, and thickness of the metallic body, as well as on the quality of its coating if present, and on the intensity of the produced anodic current.

In the case of overprotection, for instance, if steel's potential increased in the negative direction by excessive polarization in the cathodic direction, then not only oxygen would be reduced at the cathodes, but so is the hydrogen, via electrolysis reaction of water. Evolved hydrogen may cause blistering and delamination. Additionally, in environments suitable for stress corrosion, some of the hydrogen atoms produced may diffuse into the metal's crystalline structure, resulting in hydrogen embrittlement. Also, the current used for overprotection would lead to additional costs.

## 10.1 T/R Units

In impressed current cathodic protection systems, current is withdrawn from the general electricity network in the form of alternative current and inverted to direct current. A transformer-rectifier (T/R) unit is used for inversion of the alternative current (AC) to the direct current (DC). If the potential is adjusted, any desired magnitude of direct current can be withdrawn from a T/R unit, given that there is a fixed reference electrode attached to the cathodic protection system. Thus, it is possible to automatically adjust the current withdrawn from a T/R system that consists of, e.g., one mono-phase and another tri-phase T/R unit. Negative pole of the direct current produced by the transformer/rectifier (T/R) system is connected to the metal to be protected, while the positive pole is connected to the other anode present in the cathodic protection system.

Either silicon or selenium diodes are used in the rectifiers, which convert the alternative current to direct current either in the form of full or half waves. Silicon diodes are preferable, since they are 10% more efficient than selenium diodes, and they also take up less space.

An alternative current forms a sine wave due to oscillating 60 cycles per second, which is called a "ripple." This fluctuating current also reveals itself in cathodic protection, but since the variation in potential is small, e.g., between –1200 mV and –1400 mV, and since these fluctuations in potential occur in less than one in hundreds of a second, the voltmeter reads only a constant value, that is, –1300 mV, and thus in practice it does not cause any harm.

### 10.1.1 Efficiency of T/R Units

In operation, first the potential of the alternative current that is either three-phase or single-phase obtained from the electricity network is reduced down to the desired level using a transformer, then it is converted to direct current using a rectifier, usually with an efficiency of 60% to 70%.

Efficiency of a T/R unit equals to the ratio of the power of the produced direct current coming from the rectifier to the power of the alternative current generated by the electricity network:

$$\textit{power of direct current(watt)} = i(A) \times \textit{potential}(V) \tag{63}$$

and

$$\textit{power of alternative current(watt)} = \textit{electricity energy}(kW.hour)/\textit{time}(hour) \tag{64}$$

and

$$\textit{efficiency of the } T/R \textit{ unit} = \frac{\textit{power of the produced direct current}}{\textit{power of the incoming alternative current}} \times 100 \tag{65}$$

Thus, based on formulas 62 through 64, for a produced DC current of 20 A and 15 V, the power of the DC current would be 300 Watt, and if the electricity energy spent is 0.2 kW.hour for half an hour, the power would be equal to 400 Watt, resulting in a 75% efficiency for the T/R unit. Efficiency of the T/R unit increases with the potential of the withdrawn direct current. Thus, for potentials lower than 10 V, efficiency decreases below 60%. Efficiency is higher when three-phase current is used.

### 10.1.2 T/R Units with Constant Potentials

In some cases, current needed for cathodic protection changes due to appearance of stray currents due to changes in the anodic bed resistance depending on the seasonal changes; thus, the T/R unit automatically adjusts the current to keep the pipeline/terrain potential constant. For this reason, a permanent reference electrode is installed nearby the metallic structure to be protected, and the produced potential is used under the control of the current through a magnetic amplifier. If the potential is higher than the needed amount, then the amplifier stops the current for a short period of time, and if it is lower, then more current is withdrawn from the T/R unit, and hence the pipeline/terrain potential is always kept constant.

### 10.1.3 Installation of T/R Units

T/R units are installed usually on a concrete block that is at least 30 cm high or on a pole that is at least 130 cm high off the ground, so that they are not affected by the environmental factors such as rain, temperature changes, etc. Cables carrying alternative current from the electricity network should be protected by placing them into another protective cable and then clamping them onto the pole. There must be a connected switch outside of the T/R box to cut off the current. The outer box must be earthed and the cables carrying the direct current must be labeled for their positive and negative poles.

For T/R units that automatically adjust the current to maintain constant pipeline/terrain potential, a regular cable must be installed, connecting to the reference electrode near the pipeline as well.

Cables are connected via thermite welding after thorough cleaning of the pipeline surface, removing the coating until the bare steel substrate surface is exposed. At the location which is thoroughly cleaned, thermite dust is fired, the slags are removed after welding, and open areas are isolated using mastic materials.

Cables that are used to carry high intensity direct current must have large cross sections and their connections must not cause any high resistance areas, in order to prevent potential reductions. If a small hole exists on any of these cables embedded in the ground, stray currents will form, and the cable can be detached due to corrosion. A stray current will also mean that the direct current potential applied by the T/R unit to anode will decrease.

### 10.1.4 Technical Properties of T/R Units

Technical properties of a T/R unit such as its capacity, its potential and the direct current output, must be determined during the project phase based on the following criteria:

- Whether 220 V single-phase or 380 V three-phase alternative current will be used.
- Whether air-cooling or oil-cooling system will be used; T/R units with oil-cooling systems are preferred in corrosive atmospheres, while in places where there are refineries, combustible gases and petrochemicals, explosion proof T/R units with air-cooling systems must be used.
- Whether the T/R unit will have selenium or silicon diodes.

- Maximum operating temperature.
- Sensitivity and capacity of measurement devices such as ampere-meter or voltmeter that are in the T/R box.
- Where the T/R unit will be installed, whether on a pole, on the wall, or on a concrete block at the surface.

## 10.2 Types of Anodes

In impressed current cathodic protection, since energy or current is provided externally, the reference anode does not corrode instead of the metal to be protected; however, since every metal dissolves more or less with an applied potential, even the most durable supplementary anode will have a limited service life. Therefore, anodes must be economical, since half of the initial establishment expense is spent on anodes in cathodic protection systems. Thus, mass loss per current withdrawn (A.year) must be as small as possible. It is important that the current withdrawn from the anode is as highest as possible and the anode resistance does not increase in time, such as titanium anodes that are coated with conductive metal oxides, e.g., $NiO$-$Fe_2O_3$, that never passivate.

Anodes used for impressed current cathodic protection systems are placed either vertically or horizontally in anodic beds at a depth at least equaling to the length of the anode. Mostly coke dust is used as the anodic bed filling material surrounding the anode. Coke dust reduces the resistance of the anodic bed, and prevents anodic weight loss.

Issues such as location plan of the anodes, their connections, and potential losses along the connections must be addressed beforehand in impressed current cathodic protection systems. Locations where the potentials measurements will be made must be carefully selected for effective monitoring of the system.

### 10.2.1 Graphite Anodes

Graphite anodes are economical, and thus are commonly used in seawater, freshwaters, and in the ground. Since the products of the anodic reactions of graphite are all gases, no passive layer is formed at the anode surface. In the freshwaters and in the ground, oxygen and carbondioxide are produced, while in the sea, mainly chlorine gas is formed along with oxygen, especially at high current densities. It is suitable to withdraw 2.5–3.0 A/ $m^2$ current from graphite anodes, leading to a weight loss of less than 0.5 kg/A.year, while withdrawal of over 4 A/$m^2$ current in freshwaters, over 10 A/$m^2$ current in the ground, and over 30 A/$m^2$ current in the sea will cause the graphite anode to break.

Graphite anodes are placed into coke dust anodic bed material, leading to uniform withdrawal of current; however, gases formed as a result of anodic reactions must be removed from the environment, and using filtered coke dust helps. Additionally, the forming acids must be neutralized, which is usually done via addition of lime water.

An additional 1.7 V potential must be applied to cathodic protection system to make graphite the anode and to keep steel as the cathode, since the electrode potential of the graphite anode is 1.7 V, more positive than the steel. Thus, in the absence of an applied current, graphite would be the cathode and steel would be the anode.

### 10.2.2 Iron-Silicon Anodes

Cast iron anodes that are alloyed with 14.4% silicon are widely used in impressed cathodic protection systems. Silicon forms a strong silicon dioxide film on the anode surface, protecting the anode. However, this does not increase the anodic resistance, since the formed hydronium ($H^+$) ions provide the conductance despite the very high resistivity of pure silicon, as long as the surface is not entirely covered with silicon dioxide. This is achieved by alloying iron with 14.35% of silicon.

$$Si^{4+} + 4H_2O \longrightarrow Si(OH)_4 + 4H^+ \quad \text{(Eq. 90)}$$

Iron-silicon anodes are very resistant and durable in fresh waters and in the ground. Their weight loss values are around 0.5 to 0.75 kg/A.year at 20 A/$m^2$ current density. In the seawater, however, due to the formation of chlorine gas, passivation is ruptured, leading to pitting corrosion at the surface of the anode. Addition of 4.5% chromium to the alloy, making it high silicon cast iron (HSCI) alloy, prevents pitting corrosion, and thus, the weight loss reduces to 450 g/A.year at 500 A/$m^2$ current density. Addition of 3% molybdenum instead of chromium allows these anodes to be used at temperatures even above 100°C.

The biggest disadvantage of iron-silicon anodes is that they are heavy and brittle. Also, if the anodic bed dries up, the produced current or the current efficiency is substantially reduced.

### 10.2.3 Silver-Lead Anodes

Lead anodes that contain 1% to 2% silver can produce currents of 50 to 200 A/$m^2$ in the seawater. Weight loss of these anodes is about 1 kg/A.year initially, which decreases, however, down to 30 to 50 g/A.year after about a month, due to the formation of a lead peroxide layer on the surface of the anode. The lead peroxide layer prevents the anode from dissolution, but does not increase the anodic resistance, since it is conductive unless more than 400 A/$m^2$ current is withdrawn. Withdrawal of such high currents leads to the formation of nonconductive lead chloride and lead oxychlorides underneath the lead peroxide layer, increasing the anodic resistance, and thus reducing the anodic current.

The reason for addition of silver is that if lead alloy is not alloyed with silver, lead is passivated, forming lead chloride in the seawater. Silver prevents the passivation, even if sulfates are present in the environment alongside with chlorides. Increasing the silver percentage to 2% fastens the lead peroxide formation, or the same can be done via coating the surface of the anode with coke dust. However, lead peroxide layer cannot form in fresh waters and in terrains especially if the anode is operated at very low current densities.

### 10.2.4 Titanium Anodes Coated with Platinum

Titanium or niobium anodes coated with a platinum layer of 5 to 10 μm thickness make up very efficient high performance anodes, which produce very high currents in the seawater up to 1000 $A/m^2$ and 100 to 300 $A/m^2$ in terrains in coke dust anodic beds. Platinum coated titanium or niobium anodes have very little weight losses of only about 5 to 10 mg/A.year in the seawater.

Platinum coated titanium or niobium anodes are resistant to the evolution of chlorine gas; however, they are expensive, and if operated at potentials over 8 V or if there are frequency changes or ripples at T/R units due to the alternative current input, the platinum coating may be damaged, leading to the passivation of the anode.

### 10.2.5 Titanium Anodes Coated with Metal Oxides

Titanium anodes coated with metal oxides, e.g., with $NiO+Fe_2O_3$ mixture of certain proportions, are widely used since they never passivate, they have high current outputs, and their resistance does not increase in time. They can produce up to 600 $A/m^2$ current in the sea and up to 100 $A/m^2$ current in the ground in coke dust anodic bed. Coke dust surrounding the anode as a backfill material in the anodic bed decreases the resistance, thus the mass loss. Other common backfill materials used with this type of anodes are graphite, petroleum coke, and coke.

$NiO$ and $Fe_2O_3$ are coated onto the surface of these anodes via sintering so that they are not affected by the evolution of chlorine or oxygen gases at the anode surfaces, and they are also resistant to acids that have pH values as low as 1 even at very high current densities without polarization. Thus, they are especially preferred in acids and in electrolytes containing active chemicals.

Because of such inertness, weight loss of titanium anodes coated with 10% NiO and 90% $Fe_2O_3$ at 500 A/m$^2$ is 1.56 g/A.year, while it is 0.40 g/A.year for titanium anodes coated with 40% NiO and 60% $Fe_2O_3$ at 500 A/m$^2$, compared to 0.01 g/A.year weight loss of platinum coated titanium anode, 30 g/A.year weight loss of silver-lead anode, 200 g/A.year weight loss of graphite anode, and 450 g/A.year weight loss of iron-chromium-silicon anodes under the same conditions at 500 A/m$^2$ current density in the seawater.

Advantage of these anodes over platinum coated anodes is that these anodes' over potential does not rise too much due to oxygen and chlorine gas evolutions at high current densities such as at 500 A/m$^2$, since both chlorine's and oxygen's dissociation potentials are low for metal oxide coated titanium anodes. In both platinum and metal oxide coated anodes, first chlorine gas is evolved at the anodic surface; however, at high current densities such as 500 A/m$^2$, oxygen gas evolution occurs at platinum coated titanium anodes. Dissociation potentials vary very slightly with varying current densities for metal oxide coated titanium anodes. This allows high currents to be withdrawn from the titanium anodes without increasing the potential of the applied current, and as a result, making it possible to use smaller anodes, reducing the size of the anodic bed.

## 10.3 Anodic Bed Resistance

Anodic bed resistance depends on the length of the active region of the well, on the diameter of the well, on the quantity and type of the anode, on the resistivity of the terrain, and on the anodic bed filling material. The length of the inactive region of the well does not affect the anodic bed resistance. The inactive region in deep well anodic beds is the region where no anode is placed, and thus it is filled with clean gravels, which allows easy removal of the gases produced by anodic reactions that substantially increase the anodic bed resistance if not discarded.

In cathodic protection systems, resistance of the anodic bed is intended to be reduced as much as possible, since high resistances require anodic currents of higher potentials, also leading to higher costs. However, at first, current needed for cathodic protection is determined, because if a current of low intensity is going to be sufficient for the cathodic protection, anodic bed resistance may not need to be reduced, which would otherwise increase the costs. Nevertheless, if the determined current intensity requires the reduction of anodic bed resistance so that the potential of the applied current can be reduced, also reducing the overall costs, then the following measures can be taken to reduce the anodic bed resistance without reducing the cathodic protection service life:

- Coke dust is used as anodic bed filling material, increasing the effective size of the anodes.
- Anodic bed is prepared in grounds with low resistivities.
- Number of anodes is increased, which, however, will also increase the overall costs.
- Distance between the anodes that are connected parallel to one another is increased.
- A higher number of smaller anodes of the same total mass are used instead of a few big anodes.
- Anodes that have higher length/diameter (L/d) ratios are preferred.

Dwight formulas are used to measure the resistance of a single anode that has a length/diameter ratio bigger than 5 (($L/d > 5$), which has a bar shape or is cylindrical. First, half diameters of anodes that do not have circular cross sections are calculated based on formula 60. Then, a version of Dwight formula is used to calculate the anodic bed resistance of a number of anodes that are connected parallel to one another according to formula 61.

However, if the distance between the anodes is too long, such as s > 10 m, then the third ratio within the parentheses can be considered as zero, and omitted, resulting in

$$R = \frac{\rho}{2\pi L n}\left(\ln\frac{8L}{d} - 1\right) \tag{66}$$

which can be rewritten as

$$R = R_0/n \tag{67}$$

where n is the number of anodes, since resistance of a single anode that is vertically placed in an anodic bed, as stated in formula 58, is

$$R_{\text{vertical}} = \frac{\rho}{2\pi L}\left(\ln\frac{8L}{d} - 1\right)$$

In practice, having an "s" value longer than 5 m is not preferred, since it increases the costs due to more cables and excavations.

## 10.4 Types of Anodic Beds

There are two types of anodic beds used in impressed current cathodic protection systems: one is shallow well anodic bed, and the other is deep well anodic bed. In shallow well anodic beds, the anode is placed into the well either vertically or horizontally at a depth of about 2 meters then covered with coke dust, while in the case of deep well anodic beds, anodes are placed vertically at a depth of least at 15 meters into the deep well anodic bed that has a diameter of at least 20 cm, since less of a diameter would cause difficulty installing the anode. A depth of at least 15 meters is needed, since only then the increase in potential they cause at the surface would be less than 200 mV, thus not causing interference. After the anode is released into the deep well, the remaining of the well around and below the anode is filled with coke dust, while the part of the well above the level of the anode is filled with clean gravel to allow the evolved gases to escape. On the other hand, shallow well anodic beds may cause high potential changes up to 5 V at the surface.

Regularly, Werner's four electrode method is used to measure the resistivity of a soil layer at the surface of the earth. However, in the case of deep well anodic beds, the anodic bed's surroundings usually consists of many geological layers and not just soil; thus, the ground resistivity cannot be measured unless the well is opened.

Potential difference created by the anodic beds at the surface of the earth is measured by formula 49, that is:

$$\Delta E = i\rho / 2\pi r$$

where i is the current withdrawn from the anode in Amperes, $\rho$ is the ground resistivity in ohm.cm, $r$ is the distance to the anodic bed's axis, and $\Delta E$ is the potential difference in terms of Volt at a distance $r$ of from the anodic bed's axis at the surface.

Potential gradient created at the surface of the earth is in a shape of sphere, and thus lesser potential difference is created at the surface in the case of a deeper anodic bed.

Although shallow well anodic beds are less expensive and easier to install, for reasons stated below, deep well anodic beds are preferred in many cases:

1. A large surface area is needed for shallow anodic beds, which may not be possible to find, especially at urban locations or at other locations that are inconvenient, such as mountainous areas.
2. If ground resistivity is too high, use of deep well anodic beds may be of help reducing the anodic bed resistance.
3. Deep well anodic beds are not affected by seasonal climate changes or other changes happening close to the surface, such as agricultural activities or construction projects taking place at nearby areas.
4. Deep well anodic beds allow more uniform current distribution, thus creating less potential difference at the surface while not causing any interference on the surrounding metallic structures. Interference

effects of deep well anodic beds depend on the length of the anode and on the length of the inactive or inert region that is filled with gravel.

Disadvantages of the deep well anodic beds are:

1. Deep well anodic beds are about 20–25% more expensive than the shallow well anodic beds that produce equivalent currents. This cost difference decreases with increasing resistivity of the ground, and as the current withdrawn from the anode increases.
2. Anodes that wear out cannot be replaced with new ones in deep well anodic beds.
3. It is difficult to fill the deep well anodic beds entirely with coke dust, leaving holes and pores behind, which increase the anodic bed resistance. Therefore, first, the water inside the well is completely removed, before filling the well with the coke dust from bottom up. Additionally, coke dust backfill material is made more fluidic by adding about 0.5 kg of detergent to every 90 kg of coke dust, then preparing a 100 L coke dust mixture with water, which is then used to fill the well starting from the bottom of the well and up to the level of the anode.
4. Gases evolved due to the anodic reactions accumulate inside the well and cannot be easily removed; thus, a ventilation pipe is installed for their removal. This ventilation pipe must be large enough to discharge the gases and also small enough not to get clogged by the anodic bed filling material, which usually translates to a diameter size of 2 to 3 cm. Clogging in the well can be fixed by pumping pressurized air into the well using a hose placed through the ventilation. However, in time, a certain amount of the evolved gases will still accumulate, starting from the bottom of the well, increasing the anodic resistance, and thus

leading to a reduction in the produced anodic current. If none of the cited prevention measures are of any help, current provided to the anodic bed is cut off and anodic reactions are stopped. Thus, if the problem was due to an increase in pressure because of the evolved gases, a few days of wait would allow them to be discharged, resulting in a decrease in the pressure back to normal values, and if that is the case, then the anodic bed is put back into operation, this time at a lower potential for the same problem not to repeat itself. Ventilation pipe can also be used as an anode carrier used to place the anode in the center of the anodic bed. Additionally, if there is chlorine evolution as a result of the anodic reactions, connection cables of the anode must be coated with high molecular weight polyethylene (HMWPE) or ethylenechlorotriflouroethylene (ECTFE).

5. In time, anodic bed filling materials start to dry from top down, leading to a decrease in the anodic bed's permeability, increasing the anodic resistance, and thus reducing the produced anodic current. Injection of water to inside the deep well anodic bed using a hose placed through the ventilation pipe can wet the anodic bed if done in a timely manner.

## 10.5 Cable Cross-Sections

Resistances of anode connection cables must be below a certain value to prevent a decrease in the cathodic protection potential. The following formula is used to calculate the resistance of copper cables:

$$R = 0.0175\frac{L}{A} \tag{68}$$

R is the resistance of the copper cable in ohms, L is the length of the cable in meters, and A is the cross-sectional area of the cable in $mm^2$. L and A are adjusted based on the intensity of the carried current, and also on the expected cathodic protection service life.

## 10.6 Impressed Current Cathodic Protection Projects

There are a number of steps to be followed when realizing a project of an impressed current cathodic protection system:

1. There are several criteria to be determined before beginning to realize an impressed current cathodic protection project, which are:
    a. diameter of the pipeline, e.g., 40 cm
    b. thickness of the pipeline metal, e.g., 7 mm
    c. length of the pipeline, e.g., 10 km
    d. coating of the pipeline, e.g., asphalt
    e. planned service life for the cathodic protection system, e.g., 20 years
    f. terrain resistivity, e.g., 3000 ohm.cm
    g. current density based on the field experiments, e.g., 0.5 $mA/m^2$
2. Then based on these criteria, the total surface area to be protected and thereof the total current needed are calculated.
3. Consecutively, it is determined whether sacrificial anode or impressed current cathodic protection will be applied based on the total current needed and the terrain resistivity.
4. From the data available in the literature, steel pipe's inner electrical resistance per meter length is calculated given the diameter of the pipeline and the thickness of the pipeline metal from step 1.

5. Afterwards, coating resistance is calculated given the current density needed for the cathodic protection system from step 1.
6. Attenuation coefficient is calculated based on formula 68 given the steel pipe's inner electrical resistance from step 4 and given the coating resistance from step 5:

$$a = \sqrt{\frac{r}{R}} \tag{69}$$

   where r is the steel pipe's inner electrical resistance per meter length, and R is the coating resistance per meter length, and a is the attenuation coefficient.

7. Maximum length of pipeline that is insulated on both ends is determined given the attenuation coefficient from step 6 and given the potential voltages at both insulated sides.
8. Then, maximum allowed anodic bed resistance is calculated based on the literature data given the total current needed for cathodic protection from step 2. Since literature values are usually for terrains of 10000 ohm.cm, the corresponding value from the literature is multiplied with the factor obtained via formula 69:

$$F = \sqrt{\frac{r}{10000}} \tag{70}$$

   where r is the resistivity of the actual terrain in ohm.cm.

9. Consecutively, the minimum number of anodes that is required to be installed is calculated given the terrain's resistivity from step 1 and the total current needed for cathodic protection from step 2.
10. Then the required anodic mass is calculated based on formula 70, which is a revised version of formula 52, given that the service life of the anode should be at least as long as the planned cathodic protection service life from step 1, while other

parameters are based on the choice of the anode, which are available in the literature:

$$anodic\ mass(kg) = \frac{life\ of\ the\ anode(years) \times anodic\ current(A) \times anodic\ weight\ loss\left(\frac{kg}{A}.year\right)}{anodic\ current\ efficiency} \quad (71)$$

11. Resistance of the anodes is calculated based on the number of the anodes that are connected parallel to one another to the cathodic protection system, and also on the distance of the anodes from one another according to the respective Dwight formulas.
12. Anodic bed resistance is calculated depending on the resistance of the anodes and on the type of the anodic bed used, that is, whether shallow well or deep well anodic bed.
13. Then, it is determined whether smaller or bigger anodes will be installed based on the anodic bed resistance from step 11, on the minimum number of anodes needed from step 9, and on the required anodic mass from step 10.
14. Cross-sectional area of the connection cables is calculated based on the total current needed from step 2 and on the planned service life of the cathodic protection system from step 1.
15. Resistance of the cables is calculated based on formula 67 given the cross-sectional area of the cables from step 14 and the length of the cables from step 1.
16. Consecutively, voltage of the direct current produced by the transformer/rectifier unit is calculated using the following formula 71, given the total current needed for cathodic protection from step 2, resistances of the cathode from step 4 and step 5, resistance of the anode from step 12 and resistance of the cable from step 15:

$$E = i(R_{anode} + \mathrm{R}_{\mathrm{cathode}} + \mathrm{R}_{\mathrm{cable}}) + 1.7 \quad (72)$$

where i is the total current needed for cathodic protection.

17. Finally, the interference effect of the cathodic protection system on surrounding metallic structures is calculated based on the voltage of the direct current produced by the transformer/rectifier unit obtained from step 16 and on the length of the anodic bed, which depends on the type of the anodic bed used, whether deep well or shallow well, obtained from step 12.

## 10.7 Maintenance of Impressed Current Cathodic Protection Systems

Impressed current cathodic protection systems should be checked once a month for the first year the system is fully operational, then every three months afterwards.

### 10.7.1 Periodical Measurements and Controls of the Entire System

For the purpose of the maintenance, the following measurements and controls should be performed:

- Measurement of pipeline/terrain potentials at both "on" and "off" positions,
- Measurement of the direct current produced by the T/R unit,
- Measurement of potential of the direct current produced by the T/R unit,
- Measurement of the anodic bed resistance,
- Measurement of the foreign pipeline potential at both "on" and "off" positions at intersections with the pipeline that is cathodically protected,
- Control of the T/R unit in general, whether the fuses, cable connections, measurement devices, etc. are fully operational and calibrated,
- Control of insulations of insulated flanges, and
- Control of resistances of stray current connections if present.

### 10.7.2 Periodical Measurements and Controls of the T/R Unit

Apart from the periodical checks of the impressed current cathodic protection systems, T/R units must also be checked thoroughly once a year for the following issues:

- Working T/R units should make a working sound.
- T/R units are visually checked for presence of corrosion, for paint quality, and whether there is excessive heating. Corroded parts are recorded and then repainted.
- Inside of the unit is checked for presence of insects, lizards and snakes.
- Voltmeter and ampere meters are read.
- Impressed current power switch is turned off, and then all components of T/R unit are checked for presence of heat, since presence of a cold component may indicate that it is not working. Extremely hot components are also recorded.
- Measurement devices are calibrated.
- If present, air cooling ventilator and all other connections are cleaned.
- Level of the oil is checked in T/R units with oil cooling systems, and if the transformer oil color is not clear, and if it is of light color but cloudy and dark, then it is replaced.
- Cables that have broken or burnt insulations are replaced.
- Current cutters and buttons are checked and are replaced if broken and damaged.
- Impressed current power switch is turned back on and the efficiency of the T/R unit is checked via formula 64. Efficiency of the T/R unit should be around 50% to 60%, which decreases in time, and when the efficiency is lower than 25%, the entire unit must be replaced.

### 10.7.3 Commonly Encountered Problems

The following problems are the mostly encountered problems in impressed current cathodic protection systems that need to be checked for:

*i. Low Pipeline/Terrain Potential*

Presence of low pipeline/terrain potentials, despite the fact that the intensity and potential of the direct current originated from the T/R unit is sufficient, may be due to one or more of the following reasons:

1. Terrain may have become much more corrosive due to increases in underground water levels or increase in oxygen diffusion rates.
2. Insulated flanges that are used for insulating the pipes may have lost their resistance.
3. A new component may be added to the pipeline system, increasing the current needed for cathodic protection.
4. A new pipeline system may be established nearby the already existing one.
5. Pipeline coating may be damaged, leading to an increase in the cathodic protection current.

*ii. Very High Pipeline/Terrain Potentials*

Presence of either highly negative or highly positive pipeline/terrain potentials compared to the static potential, given that the intensity and potential of the direct current originated from the T/R unit is normal, may be due to one or more of the following reasons:

1. A highly negative pipeline/terrain potential may imply the presence of stray currents, originating from surrounding cathodic protection systems

or direct current sources, escaping to the metallic structure to be cathodically protected.

2. A potential more positive than the static potential may imply the presence of wrong connections, since normally the positive pole of the direct current source must be connected to the anode and the negative pole of the direct current source must be connected to the cathode, and when the cathodic protection current is applied, potential must increase in the negative direction; otherwise, as in the case of wrong connections, potential increase in the positive direction, and the pipeline becomes the anode and corrodes.

# 11

# Corrosion and Corrosion Prevention of Concrete Structures

Concrete has unique properties; it can be shaped, and its composition, and thus its mechanical properties, can be engineered as wished, which is why concrete is the most important building material in civil engineering projects. In time, concrete's relatively low tensile strength is improved substantially using steels in general, and even steel and carbon fibers in recent years; thus, strengths that were once a dream, such as 1000 $kg/cm^2$, are among the ordinary characteristics of today's regular concretes. Consequently, especially in developed countries, concrete structures are estimated to have 100 to 150 years service lives, while corrosion of the steels in the reinforced concrete remains as one of the biggest challenges against ensuring such long service lives. Corrosion of concrete steels occurs mostly due to highly permeable concretes that are of low strength as well as exceeding amounts of chlorides, low pH, and high humidity.

In the 1950s, steel inside concrete was not predicted to corrode. As a composite material, concrete will still have some pores, even it is produced using the best production method and application. Additionally, there is always some amount of water in concrete, even at the driest conditions; thus, it is always considered an electrolytic medium, even though its conductivity is very low. Thus, corrosion of reinforced steels still occurs, causing cracks in the concrete, leading to mechanical failures in aggressive media such as in marine environments, in factories where chlorides are used, and in environmentally polluted areas, in bridges and viaducts where salts are used to prevent icing, since concentrated salt solutions are formed as the ice melts, resulting in chlorides reaching the steel surface. Thus, in the 1970s, reinforced steels began to be coated with 1.5 to 2 cm. concrete, which prevents corrosion by not allowing corrosive species to reach the metal surface, due to the fact that concrete has low permeability and also by maintaining a basic medium, keeping the steel surface passivated. In the 1990s, the thickness of this concrete coating increased to 4 to 5 cm in Germany, for instance. Today, a corrosion allowance of at least 5 cm from the surface is implemented so that the start of corrosion is delayed as much as possible; however, it is usually not practical to increase the corrosion allowance too much.

## 11.1 Concrete's Chemical Composition

Concrete is not a natural material. It is a composite material mixed of cement, aggregates, and water. Cement is essentially made of lime, clay, silica, and iron oxide, and can interact with water when directly exposed to water environments or atmosphere. Once these materials are mixed in certain proportions, they are exposed to high temperatures, leading to the formation of clinker compounds, and complete placement of concrete is realized via vibrations while concrete molds are being filled. Major clinker compounds are $SiO_2$, $Al_2O_3$, and $Fe_2O_3$, which are obtained from clay's most important mineral, that is, kaolinite and limestone. Kaolinite's formula is $[Al_2Si_2O_5(OH)_4]$ or, alternatively, in terms of oxides it is $2SiO_2.Al_2O_3.2H_2O$, as it is referred in ceramics applications, while limestone is CaO, and it is obtained via the following reaction of calcium carbonate at 900°C:

$$CaCO_3 \longrightarrow CaO + CO_2 \qquad \text{(Eq. 91)}$$

Other clinker compounds that are formed from the aforementioned reagents are as follows:

- $CaO.Al_2O_3$, $CaO.SiO_2$, and in lesser amounts, CaO. $Fe_2O_3$ at 700°C to 900°C
- $5CaO.3Al_2O_3$ at 900°C to 950°C
- $2CaO.SiO_2$ at 940°C to 1200°C
- $3CaO.Al_2O_3$ and $4CaO.Fe_2O_3$ at 1200°C to 1300°C
- $3CaO.SiO_2$ at 1200°C to 1450°C

For convenience of the practitioners of the matter, these compounds are commonly symbolized with capital letters corresponding to the first letters of the first atom in the formula:

- CaO = C,
- $Al_2O_3$ = A,
- $SiO_2$ = S,
- $Fe_2O_3$ = F, and
- $H_2O$ = H

Thus, for instance, the phases within Portland cement are defined as follows:

- $3CaO.SiO_2$ as $C_3S$,
- $CaO.SiO_2$ as $C_2S$,
- $3CaO.Al_2O_3$ as $C_3A$, and
- $4CaO. Al_2O_3.Fe_2O_3$ as $C_4AF$

Consequently, Portland type 42.5 cement has the following composition of minerals: 20.04% $SiO_2$, 5.61% $Al_2O_3$, 3.27% $Fe_2O_3$, 63.01% CaO, 2.49% MgO, 2.26% $SO_3$, 0.006% Cl, 1.64% heating loss, and 1.68% undefined.

## 11.2 Corrosion Reactions of Concrete

Corrosion of the reinforced steel bar is at its maximum in the first several days of construction, while it decreases in the amount of 90% in just one month. Major corrosion reactions of steel corrosion are the corrosion reactions of iron at neutral and basic environments, since concrete provides a basic medium, which are discussed in detail in section 1.3.1, entitled "Iron, Steel and Stainless Steels," and are as follows:

Anodic reactions are:

$$Fe \longrightarrow Fe^{2+} + 2e^- \quad \text{(Eq. 92)}$$

$$Fe^{2+} + 2H_2O \longrightarrow Fe(OH)_2 + 2H^+ \quad \text{(Eq. 93)}$$

cathodic reaction is:

$$\tfrac{1}{2} O_2 + H_2O + 2e^- \longrightarrow 2OH^- \quad \text{(Eq. 94)}$$

and thus the net reaction is:

$$2Fe + \tfrac{1}{2} O_2 + H_2O \longrightarrow Fe(OH)_2 \quad \text{(Eq. 95)}$$

which becomes rust in presence of sufficient oxygen:

$$2Fe(OH)_2 + \tfrac{1}{2} O_2 + H_2O \longrightarrow 2Fe(OH)_3 \quad \text{(Eq. 96)}$$

under convenient conditions, $Fe(OH)_3$ may adhere to the surface in the form of $\gamma$-$Fe_2O_3$ and passivate it unless there are chloride ions, which react with $Fe^{2+}$ ions forming iron (II) chloride, dissolving iron from the surface into the solution, which then forms iron (II) hydroxide in the alkaline environments:

$$Fe^{2+} + 2Cl^- \longrightarrow FeCl_2 \quad \text{(Eq. 97)}$$

$$FeCl_2 + 2H_2O \longrightarrow Fe(OH)_2 + 2HCl \quad \text{(Eq. 98)}$$

HCl acidifies the anodic environment, further accelerating the corrosion process.

## 11.3 Factors Affecting Corrosion Rate in Reinforced Concrete Structures

There are many factors affecting the rate of corrosion reactions taking place inside the concrete such as pH, cement type, chlorides, temperature, and other corrosive species and gases such as sulfur dioxide.

### 11.3.1 Effect of Concrete Composition

Factors such as water/cement ratio used to make the concrete, concrete's permeability against corrosive chemicals, and its porosity are important parameters determining the corrosion rate. For example, an increase in the water/cement ratio from 0.40 to 0.60 results in approximately twofold higher diffusion rate of oxygen. Additionally, water/cement ratio should not be more than 0.45 in concretes exposed to seawater, considering the combined negative effects of chlorides along with oxygen on corrosion. Concrete can be made less permeable to corrosive chemicals by adding fine minerals to its mixing water. It can also be made less porous by adjusting the water/cement ratio and adjusting other factors associated with its preparation and curing stages.

Generally, corrosion of concrete steels for a concrete that has a corrosion allowance of 50 mm depth and a water/cement ratio of 0.60 when fully submerged in water begins in 80 days, while it begins in 380 days for a concrete that has a corrosion allowance of 75 mm depth under the same conditions. On the other hand, corrosion of concrete steels for a concrete that has a corrosion allowance of 50 mm depth and a water/cement ratio of 0.40 when fully submerged in water begins in 800 days. For concretes that are periodically exposed to wet and dry cycles, the corrosion process can start earlier.

In theory, the minimum water percentage required for hydration reactions of cement to take place is 30%; however, for economical reasons, the ratio is kept between 40% and 50% in practice, which leads to a more porous concrete due to higher water content. Air-entraining-admixtures (AEA) can be added to the concrete mixture for the concrete to have isolated pores that are not in interaction with one another, leading to a more impermeable concrete. Another reason for the porous structure of the concrete is insufficient curing of the concrete. Fresh concretes that are not cured well and are exposed to hot and dry atmospheres cannot crystallize appropriately, and thus attain a porous structure.

### 11.3.2 Effect of Oxygen

The only possible cathodic reaction at high pH values provided by the basic concrete water is the reduction of oxygen, which has to be carried to the surface of the steel embedded inside the concrete from concrete's surface along with water. However, if the concrete pores are filled with water, such as in structures fully submerged in water, oxygen must be carried via this concrete water, and solubility of oxygen in water is very low, approximately 10–13 $mg/cm^2.s$. As a result, corrosion is the highest in concretes that are exposed to periodic wet and dry cycles, due to faster ingress of oxygen as well as of chlorides. Concretes exposed to wet and dry cycles also result in regions with different humidity levels within the concrete, leading to wet regions acting as anode and dry regions, those having abundant oxygen, acting as the cathode.

Furthermore, the already low diffusion rate of oxygen in water decreases even further with increasing salt content in water, which is why corrosion rate does not increase any further over a certain percentage of salt content in aqueous solutions.

### 11.3.3 Effect of Humidity

Water is both needed as a reagent and as an electrolyte for corrosion reactions to occur. Thus, it needs to reach the surface of the steel bars embedded inside the concrete medium, which is highly possible when the concrete structure is submerged in water or when it is exposed to periodic wet and dry cycles. However, corrosion is not likely to occur in the former case, when the concrete structure is fully submerged in water, since diffusion rate of oxygen through water is very slow, and both need to be present at the steel surface to initiate corrosion.

When relative humidity exceeds 70% or critical relative humidity level is reached in polluted air, a thin liquid layer is formed on the surface of the reinforced concrete steel. Especially bridges and their foundations must be isolated well from ground humidity, so that they are not affected by deicing salts as well.

Additionally, the relative humidity levels of the environment where the concrete is cured for a few days after being casted in the mold are very important, since dry and hot climate conditions will not allow appropriate crystallization of the concrete, resulting in high amounts of pores.

If only Friedel's salt is included as the clinker compound in cement with no limestone, the hydratation reaction would be very fast. Limestone reacts with Friedel's salt ($C_3A$), forming tricalciumsulfoaluminate via the following reaction:

$$3CaO.Al_2O_3 + 3CaSO_4 + 32H_2O \longrightarrow 3CaO.\ Al_2O_3.3CaSO_4.32H_2O \quad \text{(Eq. 99)}$$

Tricalciumsulfoaluminate or Candlot's salt formed via this reaction is an insoluble compound, which precipitates over the clinker particles, forming an impermeable layer, slowing down the hydratation reaction of cement. Calcium chloride present in the concrete reacts similar to the calcium sulfate with Friedel's salt ($C_3A$), also forming an insoluble compound, although the formation is slower, and thus hydrolysis reaction of cement is not slowed down as much. Another result of the reaction between calcium chloride and Friedel's salt, other than slowing down the hydrolysis reaction of cement, is that concentration of the free chloride ions that were initially present in the concrete mix could be reduced. However, chlorides that enter the concrete externally after completion of the hydratation reactions later cannot be bound by the $C_3A$ compound.

*Concrete's Electrical Resistivity*

Another factor of corrosion, the conductivity of the concrete medium as an electrolyte, also depends on the humidity levels. Concrete's electrical resistivity is generally between 7000 ohm.cm and 100000 ohm.cm, which decreases as the concrete gets saturated with water. While resistivity of a concrete saturated with humidity is 7000 ohm.cm, it is $10^6$ohm.cm for a dry one. This reduction in resistivity can be up to hundred fold with increasing saturation levels from 20% to 80%, which is more apparent in concretes that have higher water/cement ratios. For resistivities more than 50000 ohm.cm, corrosion is usually negligible in practice, even if chlorides are present in the concrete.

### 11.3.4 Effect of Temperature

The already low diffusion rate of oxygen in water decreases even further with increasing temperatures. Further, corrosion initiation and continuation also depends on the season when it starts. Decreasing temperatures result in reducing the corrosion reaction rates, as is commonly the case for chemical reactions; however, if the temperature is below freezing point, corrosion rate theoretically becomes zero, since electrochemical reactions of corrosion require liquid water to be present as the electrolyte, which is why some cars that are left outside in icy and snowy conditions corrode less than cars left in private garages that are not aerated well. Additionally, the temperature levels of the environment where the concrete is cured for a few days after being casted in the mold is very important, since dry and hot climate conditions will not allow appropriate crystallization of the concrete, resulting in high amounts of pores.

### 11.3.5 Effect of pH

High pH levels provided by the concrete water that is basic in character give concrete one of its unique properties, that is, providing an appropriate environment for the steel bars embedded within so that they can passivate and remain passivated. Such basic character of the concrete water comes from the hydrolysis or hydration or hydratation reactions of the clinker compounds within the cement. Clinker compounds within the Portland cement hydrolyze during the making of cement based on the following reactions:

$$2\,(3CaO.SiO_2) + 6\,H_2O \longrightarrow 3CaO.2SiO_2.3H_2O + 3Ca(OH)_2 \quad \text{(Eq. 100)}$$

$$2\,(2CaO.SiO_2) + 4\,H_2O \longrightarrow 3CaO.2SiO_2.3H_2O + Ca(OH)_2 \quad \text{(Eq. 101)}$$

Produced calcium hydroxide dissolves in the concrete pore water and forms a saturated solution, increasing the pH of the concrete up to 12. Furthermore, pH of the concrete can get up to 13.2 with the hydrolysis of other alkali oxides such as $Na_2O$ and $K_2O$ that are present in the concrete in relatively low amounts. Since calcium hydroxide solution is saturated, variations in the concrete humidity do not affect the pH; however, pH decreases in time due to carbonation and alkali oxides getting washed away out of the concrete. The effect of washing away of other alkali oxides is limited to reducing the pH down to 11, which is not a sufficient factor alone to start corrosion, since iron passivates at these pH values based on Pourbaix diagrams, forming $\gamma$-$Fe_2O_3$ at the substrate surface; thus, materials made of iron and alloys embedded in concrete are passivated promptly.

*i. Effect of Carbondioxide ($CO_2$) or Carbonation*

On the other hand, the effect of carbonation has a larger impact on the pH, which can result in pH getting down to 9, which is then sufficient to inhibit passivation. Carbonation is the reaction of lime water that is present in the pores of concrete with atmospheric gases also caused by pollution such as $CO_2$, $SO_x$, or $NO_x$ via the following reaction:

$$CO_2 + Ca(OH)_2 \longrightarrow CaCO_3 + H_2O \qquad \text{(Eq. 102)}$$

When carbonation reaches the concrete coating around the reinforced steel surface, corrosion begins with diffusion of required humidity and oxygen to the steel surface, and it accelerates with diffusion of corrosive chemicals such as chlorides and carbon dioxide, leading to cracks in the concrete coating, since corrosion products are up to 10 times more voluminous than the steel alloy. However, since carbonation begins at the surface, and at the worst, 1 mm concrete/year gets carbonated, and the least corrosion allowance for reinforced concrete steel bars is commonly kept 5 cm deep from the surface, it usually does not cause any problem.

### *ii. Effect of Other Acidic Oxides*

Another reason for corrosion is if acid anhydride gases such as $CO_2$, $SO_x$, or $NO_x$ initially exist in the mixing water, aggregate, or cement, or even if they permeate the concrete after the concrete has hardened, they may reach the metal surface and initiate corrosion and especially pitting corrosion. Hence, permeability level of the concrete plays an important role, as it also does in the case of oxygen diffusion, since oxygen is needed at the surface to sustain the corrosion. As a result, more voluminous corrosion products of the acids produced by their acid anhydrides lead to increases in volume, producing structural stresses, and thus steel deteriorates and cracks occur.

Corrosive $SO_2$ gas is usually not more than 100 mg/L in most natural waters such as lakes and rivers and in waste waters; however, it may be substantially more in underground waters. In waters that have little salt concentrations, sulfates usually exist in the form of plasters, while in salty waters they exist in the form of magnesium, sodium, or potassium sulfates. Solid sulfate salts have no effect on concrete; however, when they are in aqueous solutions, they react with $CaCO_3$ and $Ca(HCO_3)_2$, producing limestone first, and then react with tricalciumaluminate, producing ettringite salt. Ettringite salt is also called Candelot salt; it is a white salt and has a needle-like crystalline structure. One mol of ettringite salt has 32 moles of hydrated $H_2O$ in its crystalline structure, and thus large increases in its volume are observed, leading to large cracks and tears in the concrete structure. Sulfates that react with tricalciumaluminate and lead to this result are usually calcium, magnesium, and ammonium sulfates.

*iii. pH Measurement*

The pH of the concrete is the pH of the concrete pore water, which is regularly saturated calcium hydroxide solution. Thus, pH measurement of a concrete is performed by a pH meter via measuring the pH of a sample that is drilled out of the concrete and grinded, which is then mixed with distilled water until attaining a muddy character, as it is the case inside the concrete.

### 11.3.6 Effect of Chlorides

Chloride ion has high electronegativity, and thus is better absorbed than hydroxide ions and oxygen, leading to the formation of iron chloride, preventing formation of iron hydroxides, which normally passivate the surface. Chlorides can break down the passive layer present on the steel surface as acids do, even at pH values over 11.5, because chloride corrosion in reinforced concretes is not dependent on pH, unlike carbonation. Iron chloride reacts with oxygen and water, regenerating the chloride ion that acts similar to a catalyst, resulting in an autocatalytic process via the following reaction:

$$3FeCl_2 + \frac{1}{2} O_2 + 3H_2O \longrightarrow Fe_3O_4 + 6H^+ + Cl^- \quad \text{(Eq. 103)}$$

Pitting corrosion occurs when these reactions occur in a narrow area. Adsorption of chloride ions also increase the potential of the steel in the negative direction, causing the steel bars that are placed close to the concrete surface to become the anode, while the steel bars that are away from the reach of chlorides become the cathode, resulting in an electron flow from the anodic regions to the cathodic regions. Reinforced concrete steels passivate at around –300 mV when standard hydrogen electrode (SHE) is taken as the reference electrode, and remain passive until 500 mV potential. However, if there is 3.5% of chloride present in concrete, then passivation of steel starts at –300 mV, but remains passive only until –100 mV.

Concrete steels are better protected from corrosion, as the corrosion allowance over the concrete steels is greater, usually between 5 cm and 7.5 cm. For concrete structures fully submerged in water, chloride concentration decreases going deep inside the concrete, while if the concrete is exposed to atmospheric air only, then the chloride concentration at the surface of the concrete will be lowest due to being washed away by rain waters.

There are two primary sources of chlorides in concrete; first is the aggregate composition or the concrete mix used in making of the concrete, and second is via diffusion from external sources such as caused by sea winds carrying chloride ions or salts used to prevent icing on the roads. Concrete that is prepared with clean and high quality materials will have chloride content to some extent, but chloride ingress causing danger is mostly due to chlorides that penetrate the concrete structure afterwards, such as salts that are used to prevent icing at highways, bridges, and viaducts, which penetrate into concrete structures, leaving chlorides behind after evaporation of water. Chlorides that originate from the concrete mix are less dangerous since some react with hydrated cement products such as tricalciumaluminate ($C_3A$), forming insoluble calcium chloroaluminates such as tricalciumaluminochloride compound, which is also called the Friedel's salt ($3CaO.Al_2O_3.CaCl_2.nH_2O$), while some are absorbed by these hydrated cement products and the rest of the chlorides remain free. Chemically bound chlorides do not take place in corrosion processes. Only the remaining free chlorides and chlorides that enter the concrete structure after hardening of the concrete from external sources cause corrosion damage in reinforced steel.

Diffusion of chlorides to the reinforced steel surface is assisted by the cations present in the concrete structure. Diffusion coefficient of chlorides combining with cations of (+2) charge is higher than with cations of (+1) charge, e.g., $MgCl_2$ is more harmful than NaCl. Additionally, diffusion of chlorides depends on the water/cement ratio and the porosity of the concrete. An increase in the water/cement ratio from 0.40 to 0.60 increases the diffusion rate of chlorides many folds from 0.01% up to 0.10%.

Especially in low pH values, fewer amounts of chlorides are sufficient to start corrosion. Thus, since harmful effects of chlorides increase with decreasing pH, $[Cl^-]/[OH^-]$ ratio is commonly used instead of the allowed maximum chloride content. For instance, for $[Cl-]/[OH-] > 5$ corresponding to 3.6 kg $Cl^-/m^3$, or in other words, %0.16 kg $Cl^-$/kg concrete at pH = 12.4, corrosion rate increases rapidly. $[Cl^-]/[OH^-]$ ratio is 3 when chlorides diffuse from external sources, and 0.6 when chlorides initially existed in the concrete mix. It is accepted that corrosion of reinforced steel begins when these ratios are exceeded. Alternatively, it is proposed that if there is more than 0.4% by weight of chlorides in the concrete structure, the service life of concrete is reduced by 80%. Supporting this claim, concrete structures in corrosive environments failed in less than 20 years.

Homogeneity of chlorides throughout the concr ete structure, and whether there are large concentration differences as well as presence of other ions such as sulfates, are among the other criteria affecting corrosion of reinforced concrete steels due to chlorides.

### *i. Chloride Content Measurement*

Due to presence of both chemically bound and free chlorides within the concrete, concrete's chloride concentration is measured in two forms, chloride content that is soluble in acids and chloride content that is soluble in water.

The depth of the concrete where the sample is drilled makes a difference in the chloride content that will be measured. Grinded concrete sample is dissolved in acid and the chloride ions are titrated with silver nitrate solution, leading to the determination of the total chloride content of the concrete that can be dissolved in acid.

In the other method, to determine the amount of chlorides soluble in water, grinded concrete sample is kept in water for 24 hours, and the chloride ions that are ionized during this period are titrated with silver nitrate in a neutral medium.

### *ii. Limit of Allowed Chlorides in Concrete*

The U.S. Highway Administration determined the limit of the total chlorides that can be allowed in regular concretes to avoid corrosion of the reinforced concrete steels as 0.20% by wt. or 4.5 kg chloride/1 $m^3$, which is the limit of the chloride content soluble in acids, of which 75% is the chlorides that can be dissolved in water, corresponding to a limit of 0.15% by wt. or 3.5 kg chloride/1 $m^3$. Corresponding limit values for prestressed concrete steels are less, with 0.08% by wt. as the limit of the chloride content soluble in acids and 0.06% by wt. as the limit of the chloride content soluble in water.

### *iii. Removal of the Concrete Chloride*

It is possible to remove some of the chlorides present in the concrete by washing the concrete with pressurized water that has a pressure of 10 MPa. Another method of chloride removal from concrete is via electrochemical ion exchange. In this method, free chloride ions in the concrete can be collected at an anode via application of an external potential. For this reason, wet sponges are attached onto both sides of the concrete, a metal plate is attached in between, and a high density current is passed such as 1–2 A/$m^2$ from this metal plate for several weeks. Chloride ions migrate towards the sponge that is used as the anode. Although this method is very effective in removal of the chloride ions, it is likely that it causes a reduction of the adhesion of the reinforced concrete steels.

### 11.3.7 Effect of Magnesium Ions

The type of the cation that the chloride ion is combined with is an important factor affecting the corrosion rate. In standards stating the corrosive chemical amounts for concrete such as BS 3148 and TS 3440, it is stated that magnesium ion concentrations between 100 to 300 mg/L would cause weak corrosion, while concentrations between 300 and 1500 mg/L cause strong corrosion, and over 1500 mg/L, very strong corrosion occurs. The pH values lower than 6.5 increases the corrosive effect of magnesium ions, while pH lower than 4.5 increases the corrosive effect of magnesium ions even more. Additionally, among magnesium salts, magnesium sulfate is more dangerous, since $Mg^{2+}$ can replace the $Ca^{2+}$ ions in calcium silica hydrates, damaging structure of the concrete via the following reactions:

$$MgCl_2 + Ca(OH)_2 \longrightarrow CaCl_2 + Mg(OH)_2 \quad \text{(Eq. 104)}$$

$$MgSO_4 + Ca(OH)_2 + 2H_2O \longrightarrow CaSO_4.2H_2O + Mg(OH)_2 \quad \text{(Eq. 105)}$$

Calcium silica hydrates are essential compounds in concrete, giving its capacity to carry high loads.

Additionally, $Mg(OH)_2$ has very low solubility, which may lead to magnesium inflatio n in concrete over long durations. Under certain circumstances, magnesium sulfate may be more corrosive than other sulfates:

$$3CaO.SiO_2 + MgSO_4.7H_2O \longrightarrow CaSO_4.2H_2O + Mg(OH)_2 + SiO_2 \quad \text{(Eq. 106)}$$

$$3(CaSO_4.2H_2O) + 3CaO.Al_2O_3 + 26H_2O \longrightarrow 3CaO.Al_2O_3.CaSO_4.32H_2O \quad \text{(Eq. 107)}$$

Formation of this large salt leads to expansions and inflation in concrete, causing the structure to crack. Due to more than wanted permeability levels, concrete becomes carbonated and loses its basicity, which would lead to steel losing its passivity, exposing it to corrosion.

## 11.4 Corrosion Measurements in Reinforced Concrete Structures

### 11.4.1 Observational Methods

Observation of stains and cracks may imply the presence of certain types of corrosion, and can be promptly interpreted; however, at this stage, the corrosion may already have propagated too much. Thus, especially for structures in seawater, observation of the first crack due to corrosion is a critical turning point and requires prompt repair. A corrosion rate of 10μ to 30μ per year would result in the concrete coating to crack within 1 to 3 years. Such a relation is simply formulated as follows:

$$\textit{time passed until first crack} = \frac{\textit{loss in reinforced steel bar diameter}}{\textit{corrosion rate}(mm/year)} \tag{73}$$

### 11.4.2 Weight Loss Measurements

Samples of the metal are exposed to very corrosive environments for accelerated corrosion to take place for certain periods of time, and after the produced corrosion products are removed using a cleaning solution such as Clarke's solution that consists of 1000 g HCl, 24 g $Sb_2O_3$, and 71.3 g $SnCl_2.2H_2O$, weight loss is determined. As a result, corrosion rate is measured as the yearly penetration distance into the metal surface in terms of cm/year via the following formula:

$$\textit{Corrosion rate}\left(\frac{cm}{year}\right) = \frac{\textit{weight loss}(g) \times \frac{24 \times 365 hours}{year}}{\textit{metal density}(g/cm^3) \times \textit{metal surface exposed to corrosion}(cm^2) \times t(\textit{experiment time in hours})} \tag{74}$$

Corrosion rate is also expressed in terms of mpy or mil/year, that is, 25.4μ/year, or in terms of percentage of weight loss in comparison to the initial weight, or as surface area loss in terms of microns. Among them, mpy is the most commonly used unit, since it takes both time and surface area into the consideration.

$$\text{Corrosion rate (mpy)} = \frac{\text{weight loss}(g) \times 3.4510^{6}}{\text{metal density}(g/cm^{3}) \times \text{metal surface exposed to corrosion}(cm^{2}) \times t(\text{experiment time in hours})} \quad (75)$$

Weight loss experiments reveal that the weight losses endured during the tests for durations of even 300 days at various chloride concentrations are negligible; however, weight loss is usually associated with uniform corrosion, and concrete may still crack due to other forms of corrosion, even if there is negligible weight loss. Thus, for a more accurate assessment, complimentary electrochemical techniques should be used.

### 11.4.3 Potential Diagrams

Determination of the chloride content of the concrete as well as determination of its potential usually based on ASTM C-876 regulations; using a copper/copper sulfate reference electrode is among the two major ways to determine the extent of corrosion of a concrete sample. In this method, corrosion potential ($E_{corr}$) is measured by employing a reference electrode that is in contact with concrete surface and the potential difference between the embedded steel component and of the reference electrode is measured using a voltmeter of high impedance. Copper/copper sulfate (CSE), $Hg/Hg_2Cl_2$ calomel (SCE), or Ag/AgCl electrode can be used as reference electrode; however, CSE electrode is unstable in solutions similar to concrete pore solutions, such as in sodium hydroxide solution, leading to potential variations of up to 150 mV in seawater, and thus the other two electrodes, and among the two, mostly saturated calomel electrode (E=0.2425V) is used.

Potential measurement is simply done via a wetted sponge that is attached to the surface that is subject to the contact, with the reference electrode right above the reinforced concrete steel. Readings obtained from the voltmeter should not fluctuate more than 20 mV both in the positive and negative directions for at least five minutes so that the readings can be taken into consideration; otherwise, the concrete surface must be wetted again and measurements must be repeated.

Based on ASTM C 876 criteria, if the measured corrosion potential or reinforced steel's half reaction potential is more negative or less than –500 mV compared to CSE electrode, or more negative or less than –350 mV compared to SCE electrode, then it is severely exposed to corrosion. If the potential is more negative or less than –350 mV compared to CSE electrode, or more negative or less than –270 mV compared to SCE electrode, then it is 90% susceptible to corrosion. Additionally, if the potential is between –350mV and –200 mV compared to SCE electrode, or between –270mV and –120mV compared to CSE electrode, then it is not certain whether there is corrosion or not. Further, if the potential is more positive than –200 mV compared to CSE electrode, or more positive than –120mV compared to SCE electrode, then it can be said with 90% reliability that the reinforced concrete steels are passivated, or there is no corrosion taking place. Lastly, if corrosion potential or reinforced steel's half reaction potential is measured more positive than CSE reference electrode, then either the concrete is dry, or the cable connection to steel is not done well, or there are stray currents originated from the surroundings. Corrosion potentials reveal qualitative information about whether corrosion will take place or not, while corrosion current densities ($i_{corr}$) can be measured to obtain quantitative information.

### 11.4.4 Polarization Curves

Polarization curves show the relation between the current density and the potential developed due to the polarization of anodic and cathodic corrosion reactions, revealing the corrosion current ($i_{corr}$) as a result. Since concrete's resistance is very high, measurements done using direct and alternative currents reveal errors, and thus methods such as electrochemical impedance spectroscopy (EIS), electrochemical noise, etc. are used, which assist the linear polarization method. If $i_{corr}$ is found to be less than 0.21 mA/cm$^2$, then no corrosion is expected, while for $i_{corr}$ values between 0.21 and 1.07 mA/cm$^2$, corrosion can be predicted to occur within 10 up to 15 years; for $i_{corr}$ values between 1.07 and 10.7 mA/cm$^2$, corrosion is predicted to occur within 2 to 10 years, and for $i_{corr}$ values more than 10.7 mA/cm$^2$, corrosion damage in the structure is expected within the next 2 years.

Studies relating weight loss test results with that of $i_{corr}$ values obtained from polarization curves reveal that $i_{corr}$ equaling to 1 A/m$^2$ corresponds to 1.16 mm/year penetration. Passivated steel has 0.1μ/year corrosion rate and $10^{-4}$ A/m$^2$ $i_{corr}$, while under the influence of chloride ions, these values are 0.1 mm/year and $10^{-1}$ A/m$^2$, respectively. $i_{corr}$ values between 0.1 and 0.8 μA/cm$^2$ are considered as negligible corrosion currents for steel components of reinforced concretes, and thus corrosion can be assumed as not occurring.

## 11.5 Corrosion Prevention of Reinforced Concrete

Among the primary methods to prevent corrosion of reinforced concrete are coating of reinforced steel with epoxies or with other protective coatings, using concretes that have low water-cement ratios, use of cathodic protection, and use of inhibitors. However, specific measures can be taken based on whether the protection is a permanent or a temporary one, and whether the building to be protected from corrosion is a new or an old one. Thus, permanent corrosion protection measures for concrete structures in general are:

- Appropriate material and design selection depending on the environment
- Use of metallic coating such as with chromates or zinc
- Use of polymeric coatings such as paints and plastic coatings

Temporary corrosion protection measures in general are:

- Use of removable oils such as paraffin preventing contact of metal surface primarily with humidity
- Use of humidity removers such as silica gel
- Use of volatile corrosion inhibitors

Consequently, to prevent corrosion in new buildings, the following specific measures are appropriate:

- Sufficient concrete coating thickness has to be planned in the projects
- Concretes of high quality that have low permeability and free of cracks must be used
- Inhibitors and curing agents must be used
- Reinforced steel must be protected via epoxy coatings and cathodic protection
- Concrete should be coated providing insulation from the corrosive environment.

To prevent corrosion in old buildings, the following specific measures are the most appropriate:

- Repairs must be performed with high quality alkali repair plasters for concretes that are cracked due to carbonation
- Use of inhibitors
- Electrochemical methods such as cathodic protection and chloride cleaning must be implemented, along with repairs and protective coatings.

### 11.5.1 Via Coatings

After hardening of the concrete, it can be coated with a monomer, which can polymerize on the concrete, resulting in an impermeable layer. Water and water vapor should be removed from the concrete pores before the application of the monomer for better adhesion of the polymer via vacuum.

Corrosion prevention for pre-stressed steel bars is done via external coating, preventing corrosive species such as oxygen, sulfate, and humidity from reaching the steel surface. Commonly applied coatings, especially for underground concrete structures, are coal tar epoxy coatings with thicknesses of 400 to 750 microns, polyurethane coatings with thicknesses of 500 to 1000 microns, and plastic coatings with thicknesses of 1000 to 2000 microns.

Coating of the hardened concrete with an admixture of Portland cement is another method of reducing the chloride ingress, although such a coating is not completely impermeable.

### 11.5.2 Via Inhibitors

There are no adverse effects of inhibitors with regard to concrete hardening. Among organic inhibitors, the ones that include sulfur, especially [$-C{=}S(NH_2)$] functional group, are found to be more effective with reinforced concretes. Inhibitors inhibit reinforced concrete corrosion via precipitation on corrosion sites, forming complex compounds or competitive adsorption on metal surfaces.

### 11.5.3 Via Cathodic Protection

Pre-stressed steel bars are also commonly protected from corrosion by impressed current cathodic protection; however, in the case of overprotection and heterogeneous distribution of the current, hydrogen gas evolves at the cathode, sometimes leading to hydrogen embrittlement due to hydrogen atoms diffusing into the metal.

# 12

# Cathodic Protection of Reinforced Concrete Steels

Cathodic protection is the most effective method among corrosion preventative measures used for reinforced concretes. However, there are some problems associated with the application of cathodic protection. The most important difficulty is related to placing anodes in the concrete. Also, it is very difficult to maintain a uniform cathodic protection current, since concrete is a very low conducting electrolyte, especially when concrete is dry. Additionally, reinforced concrete steel bars are usually attached manually and not by welding, and such weak connections make it difficult for the low potential cathodic protection current to flow through; thus, some regions may be overprotected, while some regions may not receive sufficient current needed for cathodic protection.

Despite all of the difficulties, cathodic protection is still the most effective method to prevent corrosion in structures such as bridges, viaducts, piers, parking lots, pools, pre-stressed steel pipes, etc. Regardless of the chloride content of the concrete, corrosion of the reinforced concrete steels that are underground, in the water, or exposed to atmosphere can be prevented via cathodic protection. Differences between the cathodic protection that is applied to pipelines and to the reinforced concrete steels are as follows:

1. If the connections of concrete steels are not welded during casting in molds, resistances may be formed, which cause problems in carrying low voltage currents needed for cathodic protection. Thus, electrical connections must be checked beforehand experimentally as to whether they are appropriate for cathodic protection. This can be done either by directly measuring the resistance or by measuring the potential from the concrete surface. If the difference between the potential values of two different concrete steels measured from the same location is not more than 5 mV, it is assumed that the electrical connections are sufficient.
2. Resistivity of the concrete is very high when dry. Parts of concretes that are in the atmosphere can be especially dry, while other parts can be sufficiently humid. Such variations make it difficult to determine the current needed for cathodic protection. Inaccurate determination of cathodic current can also lead to overprotection.
3. Overprotection value of normal reinforced concrete steels compared to saturated copper/copper sulfate reference electrode is –1150 mV, while it is –950 mV in the case of pre-stressed concrete steels, which is why only a narrow area can be protected with one anode. Thus, many anodes need to be used, which may lead to overprotection, causing hydrogen evolution at the cathode, which in turn can cause hydrogen embrittlement.

4. As the current passes through the cathodic protection cell, hydroxide ions increase at the cathodic regions, increasing the pH and thus assisting with the passivation of the steel; however, increas e of hydroxy ions may also lead to reactions between the alkalis and aggregates, resulting in loss of adhesion between the concrete and steel.
5. While anodes used for cathodic protection of concretes can be placed outside of the concrete structure both when the structure is underground or in water, anodes that are used for cathodic protection of concretes exposed to the atmosphere must be placed inside the concrete structure, which causes problems in practice.

## 12.1 Current Needed for Cathodic Protection of Steel Structures

Both sacrificial anode and impressed current cathodic protection systems can be applied to concrete structures; however, since concretes have very high resistivities, between 3000 ohm.cm and 20000 ohm.cm, impressed current cathodic protection systems must be applied especially to concretes that are in atmospheric conditions. Current needed for cathodic protection depends on the oxygen reduction rate at the cathode, and the amount of needed current can be reduced by painting the concrete surface, for instance, preventing oxygen diffusion. Presence of chlorides does not cause any problem for implementation of cathodic protection; however, they increase the current needed for cathodic protection, since they prevent repassivation of steel surfaces. Chloride ions migrate towards the anode in time and are adsorbed. Since diffusion of oxygen is very slow for concretes that are underground or in water, 1 to 2 $mA/m^2$ current is sufficient for cathodic protection.

## 12.2 Cathodic Protection Criteria

### 12.2.1 –770 mV Potential Criterion

Generally accepted cathodic protection potential criterion, that is, –850 mV based on copper/copper sulfate (CSE) reference electrode, is taken as –770 mV for protection of reinforced concrete steels. In other words, if potential of concrete steels are brought to –770 mV or more negative compared to copper/copper sulfate reference electrode, corrosion can be prevented. On the contrary, if the potential is more negative than –1150 mV, overprotection occurs, and hydrogen evolution begins at the cathode.

Due to the fact that potential measurements are conducted under the applied current and since the resistivity of concretes is very high, IR ohmic potential reduction must be taken into account. IR potential decrease in concretes is about 2 mV for every mm depth; thus, for concrete steels that are 5 cm deep, it comes up to about 100 mV. In other words, measured potentials are about 100 mV more negative than the real potentials.

### 12.2.2 300 mV Potential Shift

This is the difference between the potentials measured both at "on" and "off" positions. IR ohmic potential reduction must be taken into consideration for the potential measured at "on" position.

### 12.2.3 100 mV Polarization Shift

Based on this criterion, concrete steels must be cathodically polarized for at least 100 mV in the negative direction. In other words, the difference between the equilibrium potential and the potential measured right after the current is cut off should be at least 100 mV. Since both potential measurements are performed when no current is applied, IR ohmic potential reduction does not affect the results. Measurements can be performed quickly for cathodic protection systems that have been running for long periods of time, and thus the "off" potential value is read right after current is cut off, while for newly running systems, current has to be applied for at least four hours before a measurement can be conducted, since cathodic polarization can occur only after at least four hours of current application. If high relative humidity and high amounts of chlorides are present in the concrete, 100 mV polarization shift is not sufficient, and at least 150 mV polarization shift should be obtained. On the contrary, if the amount of chlorides is less than 1.2 kg $Cl^-/m^3$, then 60 mV polarization shift is sufficient.

## 12.3 Determination of Protection Potential

In potential vs. log i graph, the region where the cathodic polarization curve becomes linear yields the potential, which also yields current needed for cathodic protection. However, for the linear region to be determined accurately, external currents that have at least one hundred times higher intensity than the corrosion current must be applied, so that at least three measurements can be performed in the Tafel region. Application of such high intensity currents causes problems in practice.

## 12.4 Cathodic Protection Methods for Reinforced Concrete Steels

Cathodic protection of concrete steels is similar to that of pipelines, albeit with some differences. The following are among the commonly implemented techniques:

### 12.4.1 Via Iron-Silicon Anodes

Iron-silicon anodes have 10 to 20 years of service life, and are commonly used for cathodic protection of bridges. They are installed inside a 50 mm thick conductive coke/asphalt mixture covering the concrete surface, and there is 7.5 meters of distance between each anode. Also, some sand is included in this mixture to increase the mechanical strength, since otherwise the coating is damaged very easily. Iron-silicon anodes can only be used for cathodic protection of concretes that lay horizontally.

### 12.4.2 Via Conductive Polymeric Cage Anodes

Conductive polymeric cage anodes can be used for cathodic protection of concretes that lay both vertically and horizontally. Initially, cages made of niobium coated with platinum were used, and the upper regions of the cage were coated with mortar. The biggest problem faced in this technique was that the acids formed around the anode damaged the mortar.

Conductive polymeric anodes that have the shape of a wire cage and a size of 500 mm × 250 mm can produce 80 $mA/m^2$ cathodic protection current for 25 to 35 years.

### 12.4.3 Via Titanium Sieve Anodes Coated with Oxides

Oxide coated titanium anodes of sieve shape are made of 1 mm thick wires and have openings of 100 mm × 50 mm. It is possible to withdraw 20 mA current from 1 $m^2$ of sieve anode.

### 12.4.4 Via Conducting Paints

Conducting paints are commonly used as anodes, especially for concrete structures that are in the sea, and specifically for concrete pier poles that are at the splash zone. Conducting paints are produced by adding coke dust into acrylic polymers. This paint is soluble in water, but produces a strong coating after hardening. The thickness of this paint coating is around 400μm, and the anodes are placed over it with a distance of 3 to 5 meters between the anodes.

## 12.5 Cathodic Protection of Pre-stressed Steel Concrete Pipes

Pre-stressed steel concrete pipes are produced by coating the external surfaces of a cylindrical steel sheet with mortar of a certain thickness. After the steam curing of the concrete, external surfaces are first covered with pre-stressed steel wires, and then with mortar again on the outside. The difference between pre-stressed concretes and regular reinforced concretes is that in the case of pre-stressed concretes, the steel wires are under constant tensile stress, which results in stress corrosion, which is why pre-stressed concretes are more sensitive to chlorides.

Steel body in the inner region of the concrete is usually passivated, and thus is not affected by corrosion. Mostly, pre-stressed wires that are close to the outer regions of the concrete are affected by corrosion, and thus a good coating prevents the corrosion to a certain extent; however, it is not sufficient in terrains with low resistivities that have high salt content.

Extra care must be taken when performing impressed current cathodic protection systems, since the overprotection limit potential value for pre-stressed concrete steels is –1150 mV based on copper/copper sulfate electrode; otherwise, both stress corrosion and hydrogen embrittlement due to evolution of hydrogen gas at the cathode may occur.

Additionally, attenuation constants of bare steel wires inside the concrete are too high, and thus only a short length of pipeline can be protected from one point, which is why even the use of high potential anodes, such as magnesium or especially HP magnesium anodes, may cause overprotection at regions close to the anode; thus, in the case of sacrificial anode cathodic protection, zinc anodes are more suitable, since they have a maximum potential of –1100 mV.

# 13

# Corrosion in Petroleum Industry

Closed systems in the petrochemical industry are exposed to corrosion in general due to dissolved corrosive gases such as oxygen, carbon dioxide, and hydrogen sulfide, and also due to chloride containing aqueous phases in petroleum. A partially protective calcium carbonate layer that originated from the hardness in water and carbon dioxide does not form on the surfaces of cathodically protected metallic structures used in the petroleum industry, in contrast with other environments, where it does form due to the presence of dissolved corrosive species such as oxygen, carbon dioxide, chlorides, and hydrogen sulfide, which prevent polarization at anodic locations, thus accelerating corrosion. Instead, calcium carbonate colloidal aggregates are carried over at certain sites and precipitate, forming a non-continuous and porous structure that is not protective, even leading to galvanic cells with their surroundings causing crevice corrosion underneath the precipitates, resulting in pipeline failures.

Petroleum refineries convert crude oil into more than 2500 products, such as liquefied petroleum gas (LPG), gasoline, kerosene, jet fuel, diesel fuel, greasing oils, and raw materials used in the petrochemicals industry. During these processes, the temperature of liquid petroleum is increased up to 510°C at the heaters, and thus the relevant chemical reactions and refining processes can take place.

Corrosion gets more complex in the petroleum refining industry due to new chemicals used in modern processes, which is also reflected in petroleum prices. Various chemicals are responsible of corrosion in different stages and parts of the petroleum refining process:

- In storage tanks: $H_2S$, dissolved $O_2$, and $H_2O$.
- In pre-heating heat transfer units: HCl and sulfur compounds.
- In the tower flash zone: $H_2S$, sulfur compounds, and organic acids.
- In the tower middle zone: $H_2S$.
- In the tower top zone: HCl and $H_2O$.
- In the tower: $H_2S$, HCl and $H_2O$.
- In the vacuum oven: $H_2S$, sulfur compounds, and organic acids.
- At all stages and components: $NH_4Cl$ and sulfurs if humidified with water.
- Additionally, in boiler pipes, where there is water-steam cycle and the highest amount of heat transfer associated with boiling, many corrosion types take place:
  - pitting corrosion due to dissolved $O_2$,
  - caustic embrittlement due to high pH and high NaOH concentrations,
  - uniform corrosion due low pH, and
  - hydrogen embrittlement, abrasion, and erosion corrosion at joints and welding points.

## 13.1 Hydrochloric Acid (HCl) and Chlorides

Petroleum always has saltwater to some extent. Chlorides can also originate from the ovens in the petroleum refining unit, and also from the decomposition of organic chlorides that were initially added to hydrogenize the petroleum, which should be neutralized afterwards in order to not cause corrosion in the following stages. All chlorides are determined and reported in terms of liters of sodium chloride (NaCl) per thousand barrels, despite the fact that many other types of chlorides other than sodium chloride (NaCl) exist in petroleum. In this regard, in terms of types of chlorides and their amounts, petroleum is similar to seawater. For instance, magnesium and calcium chlorides that also exist in seawater dissolve in petroleum water, and when heated up to 149°C and 204°C (300°F–400°F), respectively, they form hydrochloric acid (HCl), which is very corrosive in aqueous solutions. Such corrosion is called low temperature corrosion, where water is in its liquid form. HCl goes through the following corrosion reactions with iron in presence of $H_2S$, which is commonly found in petroleum:

$$Fe + 2HCl \longrightarrow FeCl_2 + H_2 \quad \text{(Eq. 108)}$$

$$FeCl_2 + H_2S \longrightarrow 2HCl + FeS \quad \text{(Eq. 109)}$$

In modern refineries, petroleum first goes to chloride removal units that reduce the chloride content down to between 2.85 $g/m^3$ and 28.5 $g/m^3$ (1 to 10 lb. in 1000 barrels), but even at such concentrations, HCl corrosion still remains as an issue. Over pH 5, however, this corrosion is considered negligible. Additionally, if pH can be kept below 8, then heat transfer units and condensation pipes made of copper alloys can also be prevented from stress corrosion cracking and fatigue corrosion, especially if there is no oxygen in the environment. In other words, neutral environments that have pH in the range of 5 to 8 seem to be the most appropriate conditions for corrosion events not to occur.

High temperature corrosion in petrol refineries mainly reveals itself in the form of oxidation of steel equipments. Among the major alloying elements of steel, Al, Cr and Si provide some resistance to steel against oxidation corrosion at high temperatures, given that their percentage in the alloy is sufficient, since if their amount is little, the protection they provide is also insignificant. Other alloying elements of steel, Cu, Ni, and Sn, usually accumulate between the oxide layer and the steel surface, and thus they do not have any preventative effect on the oxidation corrosion of steel. Additionally, steels that have high C content as the alloying element undergo decarburization.

At high temperatures, atmosphere that oxidizes steel includes gases such as $N_2$, $H_2O$, $CO_2$, $H_2$, and $SO_2$, along with $O_2$. Proportions and amounts of these gases depend on air/fuel ratio, fuel composition, and temperature. $N_2$ is not involved in the oxidation process, but acts only as a diluter, diluting the atmosphere with respect to oxidizing gases. $H_2O$ and $CO_2$ are involved in the oxidation process of steel at high temperatures via the following reactions:

$$Fe + H_2O \longrightarrow Fe_2O_3, FeO, Fe_3O_4 + H_2 \quad \text{(Eq. 110)}$$

$$Fe + CO_2 \longrightarrow Fe_2O_3, FeO, Fe_3O_4 + CO \quad \text{(Eq. 111)}$$

Furthermore, produced $H_2$ and CO gases reduce the iron oxides.

When iron atoms are oxidized from the metal to the iron oxide layer at the surface, they leave pits behind, which are lessened if oxidation occurs only in presence of the noble gases along with pure oxygen or dry oxygen, compared to oxidation in presence of $H_2O$ and $CO_2$. A reason for this proposed by Rahmel and Tobolski is that $H_2O$ and $CO_2$ carry the oxygen ions from the oxide layer to the metal surface and then decompose, which results in the adsorption of these oxygen ions at the surface reacting with the metal as follows:

$$H_2O \longrightarrow H_2 + O_{adsorbed} \quad \text{(Eq. 112)}$$

$$CO_2 \longrightarrow CO + O_{adsorbed} \quad \text{(Eq. 113)}$$

Produced $H_2$ and CO gases penetrate into the oxide layer and reduce iron oxides, producing Fe ions and also $H_2O$ and $CO_2$ gases, releasing them to the oxide-atmosphere interface, where the oxidizing reaction takes place.

## 13.2 Hydrogen ($H_2$) Gas

Hydrogen atoms produced by the corrosion reactions are absorbed in the metal, which then combine and form hydrogen gas in the metallic structure that does not move but accumulates with other hydrogen molecules, causing an increase in pressure, resulting in mechanical failure and cracks. This is called hydrogen embrittlement, and is discussed in detail in section 4.2.12. Hydrogen atoms leading to hydrogen embrittlement are mainly produced by the following corrosion reaction, in the case of corrosion of steel in the absence of the protective magnetite layer:

$$3\,Fe + 4H_2O \longrightarrow Fe_3O_4 + 8H \qquad \text{(Eq. 114)}$$

In slightly basic conditions, however, these hydrogen atoms cannot reach the metal surface. Additionally, these hydrogen atoms also react with the carbon atoms in carbon steel alloy via the following reaction, which is called decarburization:

$$4H + C \longrightarrow CH_4 \qquad \text{(Eq. 115)}$$

Methane gas, similar to $H_2$ gas in hydrogen embrittlement, is accumulated at grain boundaries of the metallic alloy, and causes internal pressure.

Furthermore, steels with free carbon content that have not reacted with carburization compounds are susceptible to intergranular corrosion. This occurs when chromium carbide ($C_{23}C_6$) precipitates at the grain boundaries between 500°C and 800°C, leading to reduction of chromium concentration at locations close to the precipitation sites, reducing the percentage of chromium in the alloy below the critical concentration level of 13%, which is required for steel to maintain its passivation properties. Austenitic chromium-nickel steels can absorb more carbon with temperature, while at temperatures less than 500°C, these alloys can only dissolve a little carbon, which could lead to a carbon percentage of at most 0.02% in the steel alloy. Preventing intergranular corrosion depends on prevention of chromium carbide formation, which can either be achieved by decreasing the amount of carbon in the alloy or by adding molybdenum to the alloy, which extends the duration of thermal processing. Another prevention method involves dissolving the formed carbides at high temperatures first, followed by a rapid cooling, so that they cannot reform. Furthermore, it is also common to add other alloying elements such as titanium and niobium to the alloy, which are better carbide formers, and thus react with the free carbon content instead of the chromium, so that the chromium percentage in the alloy does not go below the critical percentage of 13%.

## 13.3 Hydrogen Sulfide ($H_2S$) and Other Sulfur Compounds

$H_2S$ causes corrosion when present along with water, producing iron sulfide scales on heated metal surfaces. If HCl is also present in the environment, then these scales are dissolved, generating $H_2S$ back. However, $H_2S$ corrosion is still very weak compared to HCl.

Pumps, oven tubes, containers, heat transfer units, towers, and components of towers are susceptible to $H_2S$ corrosion. High temperatures increase $H_2S$ corrosion substantially, especially over 232°C (450°F), because other sulfur compounds decompose to $H_2S$ at that temperature, increasing the amount of $H_2S$. If the petroleum has high sulfur content, carbon steel ovens are exposed to high corrosion rates, such as 12.5 mm (0.5 inch) thinning per year. At 482°C, however, $H_2S$ corrosion decreases, since the metal surface is coated with a coke layer.

Other than direct corrosion, $H_2S$ also causes hydrogen embrittlement, especially in catalytic decomposer and compressor units. Hydrogen embrittlement is also an issue in refining units dealing with phenol, sulfuric acid, hydrofluoric acid, and hydrocarbon mixtures.

## 13.4 Sulfuric Acid ($H_2SO_4$)

$H_2SO_4$ is used in alkylation units in concentrations of 85% to 95%. After being used, $H_2SO_4$ is either extracted with water or neutralized with caustic soda or sodium hydroxide, and sometimes both. The low initial concentration of $H_2SO_4$ due to the formation of esters is reversed when these esters decompose and form $H_2SO_4$ back. In low $H_2SO_4$ concentrations, corrosion is usually due to uniform and severe pitting corrosions.

Additionally, severe neutralization corrosion may occur at certain locations due to the elevated temperatures caused by the neutralization reactions of sulfuric acid, which can damage expensive alloys, such as Hastelloy B alloy, that has 25% to 28% molybdenum, some iron and other trace elements, with nickel making up the rest, in short periods of time. Thus, to resist such cases, and sulfuric acid corrosion in general, a special alloy, that is, alloy 20, made of 20% Cr, 30% Ni, some Cu and Mo, with stainless steel making up the rest, is developed to resist primarily for pumps and valve components. Another less expensive solution is using cast iron, which has high silica or carbon content.

Furthermore, $H_2SO_4$ is used in units separating olefins, where its concentration varies between 45% and 98%, and the temperature is between the room temperature and 121°C (250°F). $H_2SO_4$ absorbs olefins during the separation process, which are hydrolyzed afterwards, causing even more corrosion in equipments made of carbon steel, especially at welding points. Copper alloys severely corrode, as well turning into sponge-like structures.

After $H_2SO_4$ is used in such processes, unused sulfuric acid is obtained back in low concentrations. To recycle and reuse $H_2SO_4$, its concentration is increased, usually from 45% to 95%, by evaporating the water, which makes it very corrosive. Thus, carbon steel components used in the units where this concentration process is performed are coated with lead, which is also covered with bricks. Bricks are very resistant to hot acids, and have long service lives if they are well attached and if the temperature does not get over 177°C (350°F). The evaporation unit itself is made of cast iron or tantalum. Afterward, recycled concentrated $H_2SO_4$ is carried to the cooling units, and from there, it is carried to various units, where it will be reused. Carbon steel pipes that are coated with lead or carbon are used for transportation of this recycled $H_2SO_4$.

## 13.5 Hydrogen Fluoride (HF)

Some alkylation units use HF instead of $H_2SO_4$, which is more corrosive for steel when it is in concentrations of less than 65%. Monel alloy, which contains about 66 percent nickel and 31.5 percent copper, with small amounts of iron, manganese, carbon, and silicon making up the rest, is resistant to HF in the absence of water for a wide range of HF concentrations and temperatures.

## 13.6 Carbon Dioxide ($CO_2$)

$CO_2$ causes corrosion when present along with water. $CO_2$ either originates from the decomposition of bicarbonates that existed in petroleum or from the vapor used to ease the refining process. In some cases, it is also added artificially. $CO_2$ corrosion is very weak compared to both corrosions of $H_2S$ and HCl due to the weakness of acids with which it is associated, e.g., carbonic acid and bicarbonates. In condensation systems containing water vapor, $CO_2$ corrosion can be prevented simply by adding corrosion inhibitors.

## 13.7 Dissolved Oxygen ($O_2$) and Water ($H_2O$)

Both $O_2$ and $H_2O$ are carried into storage tanks via petroleum, while the carried amount may increase due to temperature changes and pumping. Dissolved $O_2$ delays cathodic polarization and humidity is usually concentrated at the ceiling and side walls, leading to corrosion. Corrosion is observed at low levels and at the ground level as well if the petroleum water contains other corrosive chemicals. If the stored product in the storage tank is a light one such as gasoline, then corrosion is usually observed in the form of pitting corrosion, and above the storage level, that is, from the mid-level of the tank and up, where wet and dry cycles are more frequent.

Oxygen corrosion may be observed in acidic, basic, and neutral environments, and severity of the oxygen corrosion increases, especially if there is flow of water at pH over 8. Oxygen induced corrosion does not depend on pH between pH 5 and 8.

Units dealing with water-steam cycles are usually made of alloys of iron, copper, chromium, nickel, aluminum, and steel. When these units are in service, corrosion is commonly due to the deposits of corrosion products containing phosphates that are carried into different units, leading to crevice corrosion. Copper alloys used in these units corrode due to dissolved oxygen, ammonia, high pH, and in high pressure environments of over 40 atm, while iron alloys used in these units corrode due to dissolved oxygen, carbon dioxide, low pH, corrosion deposits, and also in high pressure environments of over 40 atm. Thus, to prevent corrosion of both sorts, oxygen and carbon dioxide should be removed, pH should be kept at 8.8 to 9.2 range, dissolved oxygen concentration should be kept less than 5 ppm in the feeding water, and pressure is to be kept lower than 40 atm. Thus, hydrazine ($N_2H_4$) is commonly used in amounts of 50 to 200 pm, especially in turbines and in the absence of copper alloys. It is used as an oxygen scavenger to reduce the oxygen concentration, and also to maintain the pH at 10.5 together with ammonia or other ammines:

$$N_2H_4 + O_2 \longrightarrow N_2 + 2H_2O \quad \text{(Eq. 116)}$$

Additionally, inner surfaces of equipments and pipes made of iron and copper alloys are coated with black magnetite ($Fe_3O_4$) and copper(I)oxide ($Cu_2O$), respectively. Magnetite protects the steel surface underneath; however, if oxygen is abundant, then it reacts with oxygen to produce iron (III) oxide, which is not protective.

$$2Fe_3O_4 + ½\, O_2 \longrightarrow 3Fe_2O_3 \quad \text{(Eq. 117)}$$

Coating of copper(I)oxide ($Cu_2O$) can be achieved via the reaction of hydrazine with copper (II) oxide (CuO) as well:

$$4CuO + N_2H_4 \longrightarrow 2Cu_2O + N_2 + 2H_2O \quad \text{(Eq. 118)}$$

Apart from the use of hydrazine and coatings of iron and copper oxides, prevention of corrosion in water-steam cycle units and in boiler tank pipelines is commonly done via addition of inhibitors. Among the widely used inhibitors are $Na_3PO_4$ and NaOH that form protective layers, $NaNO_2$ that forms passivated layers, and $K_2CrO_4$ and $K_2Cr_2O_7$ that form both protective and passivated layers on the steel surface.

Corrosion inhibitors can be employed continuously or discontinuously, depending on the production capacity of an oil well, for instance. The practice is administration of 1 to 2 gallons of inhibitor once per week, corresponding to 25 to 50 ppm inhibitor concentration for a well that has a capacity of 100 barrels of oil production per day and especially in the beginning stages of oil extraction, while 1 to 2 gallons of inhibitor is administered twice a week, corresponding up to 100 ppm inhibitor concentration for a well that has a capacity of 200 barrels of oil production per day. Finally, for a well that has a capacity of over 300 barrels of oil production per day, continuous employment of inhibitors is more economical.

Corrosion associated with $O_2$ and $H_2O$ also takes place when the units are out of service due to repairs. The extent of corrosion taking place is different when the duration that the unit is out of service is very short, that is, up to 3 days, or short, that is, up to 3 weeks, or long, that is, more than 3 weeks. The primary corrosion reaction encountered in these cases is the dissolved oxygen corrosion producing green-black, brown, and red colored corrosion products of iron, with brown-red rust being the most common, depending on the length of the duration period.

Regularly, layers of corrosion products of iron, when it is exposed to atmosphere, consist of iron (II) oxide (FeO) on the inside followed by black magnetite ($Fe_3O_4$) and green magnetite hydrate ($Fe_3O_4.3H_2O$), respectively, with brown-red iron (III) oxide on the outside. Corrosion of iron when the units are out of service also takes place in a similar way; an air bubble attached to the iron surface enriches that part in terms of oxygen, and thus acts as the cathode, while surrounding parts on the surface act as the anode, leading to the forming iron (III) oxide to leave pits behind, causing pitting corrosion since anodic areas are larger compared to the cathode, as shown in equations 6 through 9 in section 1.3.1, "Iron, Steel and Stainless Steels." However, if there is not enough oxygen in the environment, e.g., in the storage tank, then the following partial oxidation to magnetite takes place:

$$3Fe(OH)_2 + \frac{1}{2} O_2 \longrightarrow Fe_3O_4.3H_2O \qquad \text{(Eq. 119)}$$

$$Fe_3O_4.3H_2O \longrightarrow Fe_3O_4 + 3H_2O \qquad \text{(Eq. 120)}$$

To prevent out of service corrosions, humidity is removed by drying the unit with pressurized hot air as well as by placing water dehumidifier chemicals such as silica gel and $CaCl_2$ at certain locations. In systems that cannot be dried or emptied completely, the system is pumped with $NH_3$ gas, with the condition that system does not include components made of copper or nickel alloys. As a third option, nitrogen gas is used for drying, especially at pipelines exposed to high temperatures.

## 13.8 Organic Acids

Organic acids do not cause corrosion at low temperatures; however, they are very corrosive at temperatures close to their boiling points, which may results in thinning of up to 9 mm (0.35 inch) per year, but this is reduced down to 0.6 mm (0.025 inch) per year when the temperature is reduced just 11°C (20°F) below their boiling points. In the case of aluminum, acetic, propionic, palmitic, stearic, and oleic acids cause only slight corrosion under 93.3°C (200°F), but corrosion increases substantially independent of presence of oxygen over 299°C (570°F). However, if there is oxygen present in the environment, even in the amount of 0.05% of the atmosphere, a protective aluminum oxide film can form and prevent corrosion.

## 13.9 Nitrogen ($N_2$) Compounds and Ammonia ($NH_3$)

Nitrogen compounds present in petroleum do not cause direct corrosion. Ammonia and cyanides formed due to catalytic decomposition reactions increase the pH of the environment, dissolving the protective polysulfur layer on steel surface, exposing it to corrosion, albeit in an indirect manner. However, the exception here is the copper alloys, for which ammonia causes direct uniform corrosion and stress corrosion cracking at pH levels over 8, and also at lower pH levels if dissolved oxygen is present. Uniform corrosion reveals itself with blue corrosion products of copper, which may lead to formation of mud and pollution. Although oxygen is not directly involved in the reactions, it reduces hydrogen tension on copper and maintains corrosion, and thus should be removed. Certain cyclic amines are used as buffers, and hydrazine is used to remove oxygen lower than pressures of 60 atm, at which these compounds decompose. Coating of copper alloys with copper(I)oxide ($Cu_2O$) helps prevention of corrosion due to ammonia as it does with prevention of corrosion due to dissolved oxygen, as stated in section 13.7, "Dissolved Oxygen ($O_2$) and Water ($H_2O$)."

$NH_3$ is used as a coolant and as a neutralization agent in the petroleum refining industry. Ammonia forms complexes with copper alloys at pH levels over 9.7 and concentrations of over 0.4 ppm via the following reactions:

$$2Cu + 4NH_3 + 2H_2O \longrightarrow 2Cu(NH_3)_2^+ + H_2 + 2OH^- \quad \text{(Eq. 121)}$$

$$Cu + 4NH_3 + 2H_2O \longrightarrow Cu(NH_3)_2^{+2} + H_2 + 2OH^- \quad \text{(Eq. 122)}$$

Ammonia can also react with $H_2S$, forming ammonium hydrosulfide via the following reaction:

$$NH_3 + H_2S \longrightarrow NH_4HS \quad \text{(Eq. 123)}$$

Ammonium hydrosulfide causes corrosion, especially at temperatures very close to the boiling point of water. Some of the nitrogen that exists in the burning air is converted to nitrogen oxides due to high temperatures, and may become very corrosive when combined with humidity.

## 13.10 Phenols

Phenols are used in production of aromatic hydrocarbons and greasing oils. It is sufficient to use components made of carbon steel in units dealing with phenols, where products are exposed with phenols first, and then the oil-phenol mixture is separated, and lastly, phenols are recycled in various units. Phenol induced corrosion of carbon steel increase over 204°C (400°F). Copper alloys are more resistant to phenol induced corrosion than carbon steels.

## 13.11 Phosphoric Acid ($H_3PO_4$)

Phosphoric acid ($H_3PO_4$) is commonly used in polymerization units and causes severe corrosion on carbon steels if combined with water in the form of uniform and pitting corrosion, the severity of which further increases with increasing temperature, in static solutions, and in presence of chlorides. Likewise, corrosion due to phosphates also increases in presence of chlorides as well as in presence of phosphoric acid. Phosphoric acid corrosion is the most severe during cleaning steps with water. In conditions when high water content is present, steel tubes that have diameters of 6.3 mm (¼ inch) can be punctured in as short as 8 hours. Copper alloys and Hastelloy B alloy are resistant to $H_3PO_4$ corrosion.

## 13.12 Caustic Soda (NaOH)

Caustic soda, or, in other words, sodium hydroxide (NaOH), is used to neutralize acids and also for the production of greasing oils. Carbon steels are resistant to caustic corrosion at room temperatures; however, steel components that are relieved of structural stress are susceptible to stress corrosion cracking in the form of caustic embrittlement at high temperatures. This critical temperature level can be measured using the following formula:

$$T\ (°F) = 170 - \text{density of aqueous caustic solution (Baume)} \quad (76)$$

Caustic corrosion of carbon steel begins usually over 93.3°C (200°F), and 18% Cr + 8% Ni stainless steel is susceptible to stress corrosion cracking at over 204°C (400°F). High NaOH concentrations lead to the following reaction dissolving protective $Fe_3O_4$ layer:

$$Fe_3O_4 + 6NaOH \longrightarrow 3Na_2FeO_2 + 3H_2O + \tfrac{1}{2}O_2 \quad \text{(Eq. 124)}$$

When magnetite is removed, bare iron is exposed to corrosion via the following reactions:

$$3Fe + 4H_2O \longrightarrow Fe_3O_4 + 4H_2 \quad \text{(Eq. 125)}$$

$$3Na_2FeO_2 + 4H_2O \longrightarrow Fe_3O_4 + 6NaOH + H_2 \quad \text{(Eq. 126)}$$

$$3Na_2FeO_2 + 4H_2O + \frac{1}{2} O_2 \longrightarrow Fe_3O_4 + 6NaOH + H_2O \quad \text{(Eq. 127)}$$

Magnetite that is reproduced as a result of reactions 126 and 127 is not adherent to the surface, and thus is not protective as the original magnetite layer.

## 13.13 Mercury (Hg)

Mercury is commonly used in measurement devices and in petroleum refining units. It may diffuse into different locations when equipment containing mercury is broken, leading to stress corrosion cracking in monel alloy and other copper alloys. Mercury is also very corrosive for aluminum.

## 13.14 Aluminum Chloride ($AlCl_3$)

Aluminum chloride is used in isomerization units. Corrosion due to $AlCl_3$ is negligible in the absence of water. With water, however, it hydrolyzes, forming hydrochloric acid (HCl), which is very corrosive, leading to pitting corrosion even in austenitic steels. Melted $AlCl_3$ in isomerization units is the most corrosive, which may lead up to 1.25 mm (0.05 inch) thinning in equipment made of nickel.

## 13.15 Sulfate Reducing Bacteria (SRB)

Sulfate reducing bacteria live in anaerobic environments and reduce sulfate anions with the acetic acid they produce to hydrogen sulfide, which in turn reacts with iron. Metals that are placed closely and have potential difference more than 50 mV are the most susceptible to biocorrosion due to SRB, and they corrode severely.

# 14

# Corrosion in Pipeline Systems

Use of pipelines date back to the second century B.C., when Chinese used hollowed sugar canes for water transportation, adding them end to end, while it began to be widely used only in the 1860s for crude oil transportation in the U.S. Essentially, pipelines can be defined as a system of equipments that can mobilize its contents and transport them. Pipelines mobilize the materials they carry via use of pumps if they are liquid and via use of compressors if they are gaseous and direct and protect the transportation via use of valves. Additionally, pipelines have control systems that measure the physical properties of the materials being carried, such as their flow, pressure, temperature, etc. Furthermore, pipelines have corrosion protection measures in place, protecting the pipeline from corrosion that can be caused by the environmental factors or the material that is being carried. Among other systems pipelines include are effective electromechanical scada systems that control the movement of the contents of the pipeline, which produce alarms in the case that the contents such as hydrocarbons are leaking. Scada (supervisory control and data acquisition) systems are a type of industrial control system (ICS). Overall, pipelines are decorated with high technology automation and telecommunication systems that use the power very effectively, and thus a pipeline system is a capital-high and an energy-high method of transportation. Pipelines are usually categorized based on the following criteria:

- The chemical composition of the pipeline and the way it is manufactured, e.g., whether from steel in API standards, from high density polyethylene, or from polyester with fiberglass, etc.
- Materials that are carried that is the contents of the pipeline, e.g., whether gaseous, liquid, or solid phase matters are carried.
- Nature of the contents, e.g., commercial end products, raw materials, or chemicals, etc.
- Objectives of the pipeline, e.g., whether connecting to another pipeline, used for transportation, used for distribution, etc.
- Location of the pipeline, e.g., whether it is a domestic or an international pipeline.
- The geography the pipeline is passing through, e.g., land, sea, rugged terrains, water passages, etc.
- The environment the pipeline is installed in, e.g., in corrosive soil, high resistivity terrains, etc.
- The standards to be abided with, e.g., API, BS, or ASTM standards.

## 14.1 Pipes Made of Iron and Its Alloys

Pipelines are made of different materials; cast iron pipes were used since 1815 and steel pipes were used until 1900s for water transportation. At first, none of these pipes had protective coatings on them, but their thicknesses were increased in time due to corrosion deposits. Some of these very first pipelines that were installed in dry terrains remain today. Since cast iron is more brittle and less strong compared to steel, thicker cast iron pipes were used in these pipelines. In time, this practice gave rise to the false impression that cast iron is more resistant to corrosion, while in reality all cast iron, ductile cast iron, and steel pipes have similar corrosion resistances. Essentially, different aerated terrains, presence of a coating, and the coating quality are the reasons for different corrosion resistances of pipes made of the aforementioned structural materials. Another false impression, especially among engineers, is that the cathodic protection can only be applied to steel pipelines and not to cast iron and ductile cast iron pipelines. Normally, there is no difference between these materials in terms of cathodic protection applicability; the difference is that cast iron pipes cannot be welded together, and thus usually are connected using plastic or rubber gaskets, resulting in a discontinuity in the applied cathodic protection; therefore, it is not preferred corrosion prevention technique for cast iron pipes.

### 14.1.1 Cast Iron Pipes

Cast iron pipes are manufactured from gray cast iron through centrifuging. Gray cast iron has about 3.5% to 5% carbon in the form of graphite, which is the reason why cast iron pipes are relatively more brittle. In terms of corrosion, graphite is nobler than iron, and thus it becomes the anode and iron becomes the cathode, resulting in a type of galvanic corrosion, which is specifically called graphitization. Due to accumulation of corrosion products around the graphite sheets and fibers in the structure, corrosion takes places in large areas, resulting in corrosion similar to uniform corrosion, and thus is slowed down. Thus, the fact that corrosion is slowed down in time and that it proceeds similar to uniform corrosion rather than the more dangerous unpredictable pitting corrosion makes cast iron more preferable compared to ductile cast iron and steel pipes in terms of corrosion.

### 14.1.2 Ductile Cast Iron Pipes

Ductile cast iron pipes have the same composition as cast iron pipes except for the presence of magnesium in the former. Magnesium causes the graphite in the alloy to be distributed in the form of spherical lumps, and thus the brittleness of cast iron is reduced, and its ductility is increased to levels close to the steel's.

### 14.1.3 Steel Pipes

Mild steel has around 0.1% carbon that is usually in the form of iron carbide. Steel's mechanical properties are better than those of cast iron and ductile cast iron, since silicon is also removed during the manufacturing of steel, along with the extra carbon, which is why steel can be welded and shaped easily.

Pipeline systems mainly transport hydrocarbons and water, and as such, they are categorized in the following sections:

## 14.2 Petroleum or Crude Oil Pipeline Systems

Pipeline systems carry hydrocarbons, primarily crude oil, usually from a well or a storage tank at a refinery to another storage tank, a pressure lowering station, or a measuring-control station. The most important issues associated especially with crude oil pipelines are fire hazard, occupational safety and health, environmental safety, and corrosion. Petroleum carried by the pipelines contain water, which includes corrosive chemicals such as dissolved oxygen, chlorides, bisulfate, organic acids, e.g., naphthenic acid, bacteria, e.g., sulfate reducing bacteria (SRB), and organic compounds that contain sulfur. Dissolved oxygen even in the amounts of 20 part per billion (ppb) can cause pitting corrosion in pipelines, which is also caused by dissolved $CO_2$ and $H_2S$ acting as weak acids. A petroleum pipeline system is a complete system of consecutive processes; thus, it should be considered as a whole primarily consisting of the following components:

- Initial injection stations: also known as supply or inlet stations, where usually storage facilities are located as well.
- Compressor/pump stations: the locations of these stations are based on the topography of the terrain and the type of the product being transported.
- Partial delivery stations or intermediate stations: where part of the product is separated and allowed to be transported elsewhere.
- Block valve stations: can isolate any segment of the pipeline for maintenance work or isolate a rupture or leak. Block valve stations are usually located every 20 to 30 miles (48 km), depending on the type of pipeline. The locations of block valve stations depend on the nature of the product being transported and the trajectory of the pipeline.
- Pig-launching stations: where various maintenance operations such as cleaning and inspecting of the pipeline are performed without stopping the flow of the product.
- Regular stations: where some of the pressure can be released. Regular stations are usually located at the downhill side of a peak.
- Final delivery stations: also known as outlet stations or terminals, where the product is distributed to the consumer. Final delivery stations could be the tank terminals for liquid pipelines or connection to a distribution network in the case of gas pipelines.
- Recycling units.
- Storage units.
- Refining units.
- Loading units, etc.

## 14.3 Water Pipeline Systems

The very first water pipelines used in the Middle Ages were made of ceramics, while the second generation of pipelines was of lead at first, but was replaced, however, with galvanized pipelines due to the lead being hazardous. In the last decades, galvanized pipelines are also being replaced mostly with copper pipelines, which are also susceptible to corrosion. Thus, plastic pipelines that are increasingly being used are considered as the third generation of pipelines, coming after pipelines based on ceramics and metals.

Water pipelines embedded in walls or underground cannot get passivated when they are not fully covered, and thus act as anodes in presence of a little amount of humidity, leading to corrosion. Corrosion increases with the increasing ratios of cathodic surface areas to anodic surface areas. Corrosive species such as chlorides, ammonia, sulfates, and nitrates exist in concrete aggregates, alums, plasters, and wood preserving paints. Humidity exists in thermal isolation systems, such as in pores of mineral wool and foam, eventually reaching pipelines.

### 14.3.1 Water Pipelines Made of Iron and Steel

Stainless steel pipelines are also used in water systems, especially in boilers and heating bars, where other materials cannot be used. The best stainless steels with regards to corrosion resistance used in pipelines contain at least 16% chromium, 9% nickel, and also some molybdenum. However, DIN 50930 guidelines state that pipelines made of iron and alloys of iron are not suitable to be used as structural materials in water pipelines due to their high corrosion susceptibilities, and also due to their corrosion products mixing into water.

### 14.3.2 Galvanized Water Pipelines

Galvanized pipelines are preferred over iron and steel pipelines for transportation of water. In galvanized pipelines, zinc mixes into water, and in time, as the protective zinc layer corrodes away, iron and alloys underneath are exposed, which then also results in formation of rust and lime deposits. Thus, galvanized pipelines should not be used in following conditions:

1. In waters rich in carbon dioxide, that thus have pH values lower than 7.3.
2. After copper and copper alloys are used, since dissolved copper ions possess high corrosion risks. If galvanized pipelines must be used after copper alloys, they should be installed in the flow direction. However, partial replacement of copper pipelines with those of galvanized ones leads to galvanic corrosion as well.
3. Over 60°C, because over 60°C, potential of galvanize exceeds that of iron's, reversing the zinc's protective effect, making iron, the anode, and zinc that is in galvanized coating, the cathode. Further, today, galvanized pipelines cannot be used for hot water transportation anymore, because cleaning of the legion microbe, which causes legion disease, requires temperatures of over 60°C.

However, if galvanized pipelines are produced and installed according to DIN 50930 norms, their service life extend to 50 years.

### 14.3.3 Water Pipelines Made of Copper

Pipelines of copper and alloys are resistant to corrosion if a protective copper oxide layer can form that contains green colored copper stone (malachite) or basic copper carbonate, $Cu_2CO_3(OH)_2$. Copper mixing into water is not harmful, as long as pH is over 7. Copper pipelines are susceptible to pitting corrosion, both in hot and cold water systems. In cold waters in less than 40°C, pitting corrosion is detectable due to more than usual green malachite formation, while in hot waters, there is no such outcome of pitting corrosion. Pitting corrosion in cold waters are observed more in underground systems than in others, possibly due to the presence of organic species inhibiting corrosion to some extent. Even chemical conditioning of water with inhibitors such as phosphates and silicates cannot sufficiently prevent corrosion of copper pipelines in cold waters. On the other hand, pitting corrosion of copper pipelines in hot waters, which is mostly seen in waters with high carbon dioxide content, can be prevented by chemical conditioning with alkali phosphates and silicates. Another type of corrosion seen in copper pipelines is erosion corrosion, occurring mostly in parts of the pipeline directly exposed to water, since the protective copper oxide layer is removed by the fast flowing water.

### 14.3.4 Water Pipelines Made of Brass

In brass pipelines, the amount of zinc decreases with increasing corrosion or in other words, dezincification, as the yellow color of zinc exposes red colored copper. Dezinfication occurs in both hot and soft waters that have pH over 8.3 and also chlorides. Chemical conditioning of water does not prevent corrosion of brass; instead, brass should be rich in copper in amounts of more than 65% in order not to undergo corrosion.

# 15

# Cathodic Protection of Pipeline Systems

Cathodic protection of pipelines was first realized in 1928 for a crude oil pipeline system by R. J. Kuhn in the United States. Before cathodic protection is administered to a crude oil pipeline system, some previous exploration work must be done, including determination of passages through water bodies, neighboring railway systems that work with direct current, high voltage transmission lines, and other pipeline systems in the surroundings. If such other structures are present, then specific measures must be taken beforehand to prevent interference corrosion. Besides, the type and the resistivity of the terrain the pipeline will pass through, underground water levels, pH levels, and redox potentials that indicate the corrosivity of the terrain are among other criteria to be determined beforehand.

## 15.1 Measurement of Terrain's Resistivity

Resistivity of the terrain where the pipeline passes through is measured in every 100 to 200 meters of the pipeline. This range can be increased to 500 to 1000 meters if the terrain is known to be same throughout that distance. If the difference in the consecutive measurements comes out very large such as twice as much or less, then more measurements must be conducted within that distance. Resistivity of the terrain is measured directly via Werner's four electrode method in the field. In this method, four electrodes are installed in the terrain equally apart from one another. Afterwards, alternative current is applied to the terrain via the two electrodes that are the most apart from each another. The potential difference that the applied alternative current produces in the terrain is measured by the other two electrodes that are on the inside, while resistance is measured via Ohm's law. Consequently, resistivity is calculated using the following formula:

$$P = 2\pi aR \quad (77)$$

where P is in ohm.cm, a is the distance between the two inner electrodes in cm, and R is the resistance read from the device in ohms. Resistivity measurements are usually performed at specific locations of the terrain that remain right above the pipeline and close to the surface. In practice, this distance is usually 150 to 200 cm. Thus, if "a" is taken as 160 cm, for instance, in the formula, "$2\pi a$" term would be approximately 1000, and 1000 times the read resistance would be equal to the terrain's resistivity, as it is commonly expressed by practitioners of different backgrounds work in the field:

$$P \sim 1000R \quad (78)$$

Terrain resistivity measurements are also performed in the laboratories using the method of rectangular prism boxes by performing tests on different samples taken from the terrain; however, errors should be expected, since there are usually other factors in the natural environment that are not taken into account in laboratories. The following formula is used when rectangular prism boxes are used to determine the resistivities:

$$\text{P} = R\frac{W.L}{D} \tag{79}$$

where P is terrain's resistivity in ohm.cm, R is the measured resistance of the terrain sample in ohms, D is the distance in cm between the points where the samples are taken from, and W and D are the width and length of the rectangle, respectively.

Terrain resistivity increases substantially when humidity levels drop below 5% and decreases substantially when the humidity level reaches 20%.

Additionally, terrain resistivity increases in general with decreasing temperatures and increases substantially below 0°C, since water freezes. Thus, temperatures must be recorded during terrain resistivity measurements. The following formula is used to calculate the changes in terrain resistivity with respect to the changes in the temperature:

$$\text{P}_{T_1} = \text{P}_{T_2}\frac{40}{25 + T_1} \tag{80}$$

where $\text{P}_{T_1}$ is the terrain resistivity in ohm.cm at temperature $T_1$, and $\text{P}_{T_2}$ is the terrain resistivity in ohm.cm at temperature $T_2$.

## 15.2 Potential Measurements

### 15.2.1 Redox Potential of the Terrain

Redox potential indicates whether the terrain is corrosive and whether there is the danger of microbial corrosion. Potentials less than 100 mV indicate the possibility of severe corrosion, potentials between 100 and 200 mV indicate the possibility of high corrosion, potentials between 200 and 400 mV indicate the possibility of moderate corrosion, and potentials higher than

400 mV indicate the possibility of mild corrosion. Redox potential of a terrain increases with increasing amounts of oxidizing agents such as the dissolved oxygen in the terrain. Redox potentials are measured via the following formula:

$$E_{redox} = E_p + E_{ref} + 60(pH - 7) \quad (81)$$

$E_{redox}$ is the terrain's redox potential in mV, $E_p$ is the measured potential of platinum electrode installed in the terrain in mV, and $E_{ref}$ is the potential of the reference electrode compared to standard hydrogen electrode (SHE) in mV, e.g., for saturated $Cu/CuSO_4$ (CSE) electrode, this potential difference is 316 mV.

### 15.2.2 Static Potential & On-Off Potentials

Static potential ($E_s$) is the potential of the pipeline system measured before cathodic protection is administered, while on potential ($E_{on}$) is the potential measured under the cathodic protection current, and off potential ($E_{off}$) is the potential measured about 5 seconds after stopping the cathodic protection current. Measurements of these potentials are crucial to see whether the criteria mentioned in section 8.7, "Cathodic Protection Criteria," are satisfied, which establish whether there is sufficient polarization. Potential of the steel structure must be sufficiently polarized so that it can be cathodically protected. The following formulas are used to calculate $E_s$, $E_{on}$, and $E_{off}$ potentials.

$$E_{on} = E_s + E_p + E_{IR} \quad (82)$$

and

$$E_{off} = E_s + E_p \quad (83)$$

$E_{IR}$ becomes zero during the measurement of $E_{off}$ potential due to cessation of the applied cathodic protection current, while another potential value originated by the cathodic protection current that is the polarization potential ($E_p$) does not become zero, and decreases gradually after the cessation of the current, and thus can be measured. Based on the measured values of static potential ($E_s$), on potential ($E_{on}$) and off potential ($E_{off}$), potential shift, and the polarization shift are also calculated as follows:

$$potential\ shift = E_{on} - E_s \tag{84}$$

and

$$polarization\ shift = E_{off} - E_s \tag{85}$$

### 15.2.3 Measurement of Pipeline/Terrain Potential

Potential of the pipeline/terrain system is measured using a voltmeter or a potentiometer of a high resistance. The negative pole of the voltmeter is connected to the pipeline and the positive pole is connected to the reference electrode. Optimum measurement can be done by welding or soldering the cable to the pipeline, which is not usually practical in the field; instead, first, the rust layer is removed from the surface of the pipeline where the measurement will be conducted, and secondly, the experiment is performed by pressing the sharp metal tip of the connection cable onto the cleaned surface. The reference electrode must also be very close to the pipeline surface to prevent the IR ohmic reduction in potential, which is also not practical to achieve in the field, since the pipeline is underground, and thus usually the reference electrode is placed right above the pipeline at a hole dug at the surface. Furthermore, if the terrain is dry, it is wetted to reduce the resistance.

## 15.3 Determination of Coating Failures Based on Potential Measurements

### 15.3.1 Determination of Coating Failures Based on the Measured Pipeline/Terrain Potentials

Pipeline metal is negatively charged and the terrain is positively charged when the pipeline/terrain system is considered. As the pipeline passes through different terrains, pipeline/terrain potential varies depending on the type of the terrain and its structure. Locations where pipeline/terrain potentials are more negative become the anode and locations where pipeline/terrain potentials are more positive become the cathode, resulting in the current to flow from the anode to the cathode. Locations where there are coating failures would have more negative potentials, and thus, by measuring the pipeline/terrain potentials along the pipeline, corrosion zones can be determined.

Pipeline/terrain potentials of the locations where there are coating failures on the pipeline increase in the positive direction after cathodic protection is applied. For this reason, current produced by the anodic bed of the cathodic protection system enters the pipeline from the locations where there are coating failures, thus decreasing the potential.

### 15.3.2 Determination of Coating Failures Based on the Pearson Method

The Pearson method is commonly used to automatically determine the coating failures on a pipeline system. The Pearson method is based on monitoring the potential variations that occur due to coating failures when a high voltage alternative current is applied to the pipeline system.

The Pearson method can also be applied to pipelines that are under the cathodic protection current with the conditions that direct current, instead of alternative current, and also a different amplifier must be used. Furthermore, today, motored devices have been developed that move alongside the pipeline and detect the signals.

## 15.4 Measuring Potential along the Pipeline

### 15.4.1 Long Cable Method

Potential can be measured in equal distances along the pipeline via the long cable method, using a sufficiently long cable that has a thick cross-section. The cable's resistance can be neglected when realizing the long cable method. Potential of up to a 1 km portion of the pipeline can be measured via this method, if there are no structures within that distance that may be of any interruption.

### 15.4.2 Double Electrode Method

Another method used to measure the potential along a pipeline is the double electrode method, since it is usually not possible to install a cable of sufficient length along the pipeline, due to the presence of various structures in the path; thus, potential differences are measured alongside the pipeline using two reference electrodes.

## 15.5 Maintenance of Pipeline Cathodic Protection Systems

Cathodic protection systems used for protection of pipeline systems from corrosion are usually predicted to have a service life of 10 to 15 years. Cathodic protection systems must be checked at least once a year even if there is no failure or a breakdown for the following issues:

- Pipeline/terrain potential must be checked in every measurement station along the pipeline.
- In areas where there is suspicion of corrosion, additional potential measurements should be performed via double electrode or long cable method.
- Coating resistance must be checked along the pipeline, and at locations where there is coating damage, additional measures must be taken.
- Current efficiencies of T/R units must be checked.
- In the case of impressed current cathodic protection systems, anodic bed resistances must be checked.
- In the case of sacrificial anode cathodic protection systems, anodic potentials and the intensities

of currents withdrawn from the anodes must be checked.

- At locations where two pipelines meet, current intensity measurements must be performed both at "on" and "off" positions, under the applied cathodic protection current and when it is stopped, respectively.
- At locations where there are profile pipes passing by, the profile pipes' and pipeline's potentials should be checked, and it has to be determined whether there is contact between these metallic structures or not.
- Parts of the protected pipeline and the foreign pipeline or another metallic structure that have interference effects must be checked.
- Insulation quality of the insulated flanges must be checked via measuring their resistance.
- Surroundings of the pipeline should be checked for structures that were not initially present such as high voltage transmission lines, railway systems that work with direct current, and other foreign metallic structures.

## 15.6 Measurement Stations

Regardless of the type of the cathodic protection system used, pipeline systems must have measurement stations alongside the pipeline. These measurement stations must be installed in a way so they are easily accessible, easily turned on and off, and resistant against external factors. An altitude of 1 meter high over the ground is appropriate for their installation, which may not always be possible to realize, however, at urban locations; thus then they are either installed on a concrete block, or they are embedded in the ground. Measurement stations primarily determine pipeline/terrain potential, stray currents, and corrosion due to coating failures, and they consist of sub-units that deal with the following measurements and controls:

1. Measurement of the regular pipeline/terrain potentials via STP units
2. Measurement of the current intensity flowing over the pipeline via ATP units
3. Measurement of the current and potential at galvanic anodic connections via SATP units
4. Control of the insulated flange insulations via SIF units
5. Control of the stray currents at intersections via EPC units
6. Control of the contacts at profile pipes passing by via CTP units

### 15.6.1 STP Regular Measurement Units

STP regular measurement units are installed at periodical intervals alongside the pipeline, which are connected to the pipeline via two cables that are side by side. Although it is possible to measure the pipeline/terrain potential using only one of the cables, the second cable is installed; hence, if current needs to be applied through one of the cables, then the other can be used to measure the pipeline/terrain potential.

### 15.6.2 ATP Current Measurement Units

ATP units measure the current intensity flowing over the pipeline. For this reason, two cable connections are made to the pipeline, which are 30 meters apart. Two more cable connections are made that are 10 cm away each from the two previous locations, to apply current. One of the cables is chosen in black color and the other is chosen in white color so that current flow direction can be determined.

### 15.6.3 SATP Galvanic Anode Measurement Units

Galvanic anodes are not directly connected to the pipeline system, but over a (SATP) galvanic anode measurement unit that has two cables that are connected to the pipeline. One of these cables is connected to the cable coming from the anode and when needed, a resistance may be placed in between, and the current withdrawn from the anode can be adjusted, while the other cable is used to measure the pipeline/terrain potential at both "on" and "off" positions, when the anodes are operational and when they are not, respectively.

### 15.6.4 SIF Insulated Flange Measurement Units

SIF insulated flange measurement units check the insulation of the flanges, and thus are installed at locations where there are insulated flanges. For this reason, two sets of two cables are connected to both sides of the flange in 10 cm distances, with one set having white and the other set having black color. When the resistance of the insulated flange is measured, out of the four cables, two cables that are on the outside are used to apply the current, and the inner two cables are used to measure the potential difference.

### 15.6.5 EPC Equivalent Potential Measurement Units

EPC equivalent potential measurement units control the stray currents at intersection points of two different pipeline systems. Two sets of two cables are connected to each pipeline system with each set having a different color, of which two are used to measure the potential difference, while the other two, which have thicker cross-sections than the ones used for potential measurements, are used to place an interference connection resistance between the pipelines to control the stray currents.

### 15.6.6 CTP Measurement Units

(CTP) measurement units measure whether there are contacts with the profile pipes that are in the surroundings and are passing by. For this reason, two cables are connected to the pipeline and one to the profile pipe.

## 15.7 Static Electricity and Its Prevention

In addition to the interference effects of pipeline cathodic protection systems that are discussed in detail in section 8.9, "Interference Effects of Cathodic Protection Systems," there is also the issue of static electricity, which needs to be prevented. Static electricity occurs when materials that have very low conductance such as petroleum derivatives and hydrocarbons are carried via pipelines. Static electricity may get accumulated on the pipeline metal, and may rise up to thousands of volts and a few miliamperes based on the flow rate. This static charge is spontaneously, but very slowly, discharged to the atmosphere and to the earth.

For the static charge to become practically harmful, turbulence must be present, and the flow rate and the amount of fluids carried must be over a limit. Additionally, the contact surface of the fluid and the inner surface of pipeline metal must be sufficiently large. Resistivity of the fluid is another important criterion, since if it is higher than $5.10^{12}$ ohm.cm, the rate of discharge of the static electricity would be slower than its rate of getting accumulated on the pipeline; thus, a substantial amount of static electricity may get accumulated. Light petroleum derivatives have resistivities of around $10^{11}$ ohm.cm, resulting in static electricity being accumulated in less than 1 second.

As the fluids are entering the pipeline, they have equal number of positive and negative charges; however, in case of high flow rates, the negative charges of the fluid are absorbed by the more conductive metal ions of the pipeline, resulting in the fluid to attain plus charges and the pipeline to attain negative charges. If the coating of the pipeline is not good, then since the accumulated electricity would be discharged into the earth, potential of the static electricity on the pipeline cannot be over a certain value, and below 1500 Volts, discharges do not occur in the form of sparks. However, at higher potentials, sparks may occur with a risk of fire, and the pipe can be punctured. The static electricity may reach the tank if no insulated flange exists between the location where the static electricity gets accumulated and the storage tank. On the other hand, an explosion-proof surge protector is added to the location where there is insulated flange.

Especially in stations where storage tanks on land are loaded with petroleum products from ships, and also in pipeline systems carrying petroleum products, flow rate must not be over 0.9 m/s, and thus a pipeline that has a diameter of 10 cm can carry at most 2.27 $m^3$/min. fluids. Static electricity can get accumulated on petroleum derivates while being carried in ships due to vibrations and shaking as well. Static electricity must be earthed when the petroleum derivates are being unloaded from ships so that it is not carried over to the storage tank, which can be achieved via an insulating flange that is placed on the pipe, connecting the ship to the storage tank.

## 15.8 Cathodic Protection of Airport Fuel Distribution Lines

Sacrificial anode cathodic protection is theoretically appropriate for cathodic protection of airport fuel distribution lines for both economical and technical reasons; however, the following problems are commonly encountered in practice:

- It is very difficult to find available places at airports to install galvanic anodes.
- Measurement units for anodes cannot be installed.
- It is very difficult to manage maintenance operations, especially during a failure.

Therefore, despite the fact that it is less economical, deep well anodic beds are used, and impressed current cathodic protection systems are preferred in general.

## 15.9 Cathodic Protection of Water Pipelines

To prevent corrosion in water pipelines using cathodic protection, direct current is applied to pipelines with sufficient magnitude and potential. Sufficient pipeline-ground potential difference is –850mV at the minimum, while it is –1200 mV at the maximum in the case of polyethylene (PE) and polyvinylchloride (PVC) coated pipelines, and it is –2500mV for pipelines coated with bitumen compared to copper/copper sulfate (CSE) electrode at every point of the pipeline. However, if pipeline is not buried in the ground, pipeline-ground potential measurements may be inaccurate, and thus should not be measured and relied on unless there is a need. If applied current becomes a stray current, for instance, due to the anode bed being placed close to a railway, then cathodic protection may become ineffective. To prevent such a problem, the location of the anode bed may be changed, or its type may be changed from a horizontal anode bed to a vertical anode bed. Additionally, anodes of more resistant alloys, such as titanium anodes coated with composite metal oxides, can be used.

# 16

# Corrosion and Cathodic Protection of Crude Oil or Petroleum Storage Tanks

Petroleum storage tanks carry heavy loads and store very high volumes of petroleum. They are exposed to humidity and corrosive chemicals, especially when they are located nearby the sea, resulting in corrosion. Corrosion in petroleum storage tanks occasionally lead to leakages, resulting in loss of petroleum products, especially in underground tanks that are older than 10 years and have no corrosion protection measures. In new tanks, pitting corrosion is the more commonly observed types of corrosion, compared to other types, while in older tanks, uniform corrosion is the major corrosion type, since, in time, anodic and cathodic areas get close to each other in terms of areas they cover inside the tanks.

Petroleum storage tanks are exposed to corrosion either from underneath, where it is in contact with the accumulated rain and with ground waters, or from within the tank, especially due to the seawater present inside the tank transported, along with the material being stored such as petroleum. Corrosion of petroleum storage tanks from underneath due to contact with the accumulated rain and with ground waters is commonly prevented via impressed current cathodic protection, which does not work in the case of iron sheets that compose the lower portion of the tank, since they are not in contact with the ground; thus, when the tank is either entirely full or empty, iron sheets elastically deform, resulting in gaps between the storage tank and the ground. Corrosion of petroleum storage tanks from underneath due to contact with the accumulated rain and with ground waters is also prevented using coatings; however, they degrade in time, leading to accelerated corrosion from the locations, where there are coating failures, which occur commonly due to variations in tank loads. Thus, even if one or more corrosion prevention techniques are employed, corrosion of petroleum storage tanks has to be still monitored.

Secondly, the liquids stored inside the storage tanks, such as petroleum, also cause corrosion. When the most commonly stored liquids in petroleum storage tanks are compared in terms of corrosivity, diesel fuel is the most corrosive, followed by radiator fuel, unleaded gasoline, fuel oil, regular gasoline, jet fuel, premium gasoline, and kerosene, in descending order. However, slight differences between these different liquids in terms of induced corrosion imply that the majority of the corrosion is caused by the seawater and dissolved salts such as chlorides, because despite stored petroleum liquids having different corrosivities, none can cause significant corrosion alone, since they are not good electrolytes.

## 16.1 Cathodic Protection of Inner Surfaces of Crude Oil Storage Tanks

Overground tanks made of cylindrical steels are commonly used for storage of crude oil in the industry. Lower surfaces of these tanks are coated with a proper coating during the manufacturing stage for corrosion prevention; however, corrosion still occurs in oil storage tanks due to corrosive liquids accumulating at the lower phase, which contain concentrated dissolved salts and organic acids, along with the humidity of the tank leading to leakages and holes in time at the base of the tank, which can go out of control. Corrosion occurs mostly in the form of pitting corrosion due to heavy loads, due to tensions caused by the alternating cycles of loading and emptying processes, and also due to deformations occurring commonly at the bases of these tanks in 5 to 10 years.

For the damaged coating to be repaired, crude oil storage tanks are emptied every 5 to 10 years to undergo maintenance, where the damaged paint and the coating are removed, which all cause difficulties, resulting in the tank being out of service for long periods of time. Further, during maintenance, leakages and punctures due to corrosion are repaired, and corrosion deposits are cleaned. Old paint is removed using sand spray, since the surfaces must be cleaned near white level (SP-10) so that the coating of a new paint or fiberglass reinforced plastic (FRP) coating can be durable. The process of renovation of the storage tank coating, especially in the old, deformed crude oil storage tanks, costs a great deal of hard work and money. Hence, cathodic protection can be more economical than recoating of the paint.

### 16.1.1 Corrosion Prevention

The base that the tank is placed over is specially prepared for better protection, using crushed stones at the bottom that are squeezed with clean sand and with concrete asphalt on top. Even very strong coatings of fiberglass plastics are not the best solutions, since they require very thorough cleaning of the surface, which requires the tank to be out of service for very long periods of time during maintenance. Therefore, along with a proper coating, sacrificial anode cathodic protection is the best corrosion prevention method for crude oil storage tanks. Impressed current cathodic protection is usually not preferred due to fire hazard risks and also due to overprotection resulting in coating damages.

### 16.1.2 Sacrificial Anode Cathodic Protection

Indium alloyed aluminum anodes are preferred for galvanic anode cathodic protection of petroleum storage tanks. The biggest obstacle against the use of cathodic protection technique as a corrosion prevention method is the oxide layer covering the tank's surface, which has to be removed using a high potential anode. Thus, a dual sacrificial anode system using magnesium and aluminum is appropriate. In this method, anodes are installed parallel to one another, and magnesium anode that has the higher potential starts working soon after the application of cathodic protection current, promptly polarizing the inner bottom surface of the tank cathodically, which is covered by oxide. Thus, most of the applied cathodic protection current is used to reduce the $\gamma$-$Fe_2O_3$ layer present at the inner surface of the tank. After the magnesium anode goes flat, reducing the oxide layer, aluminum anodes come into place, providing the current needed to maintain the potential of the cathodically polarized surfaces.

In dual galvanic anode systems, a combined potential is formed soon after cathodic protection current is applied, with around –800 mV coming from magnesium anodes and –70 mA from aluminum anodes, totaling to –870 mV. The magnitude of this combined potential depends on the following oxidation and reduction reactions of the anode and cathode:

Anodic reactions:

$$Mg \longrightarrow Mg^{2+} + 2e^- \quad \text{(Eq. 128)}$$

$$Al \longrightarrow Al^{3+} + 3e^- \quad \text{(Eq. 129)}$$

Cathodic reactions that are also stated in Eq. 29 and Eq.77

$$2H^+ + 2e^- \longrightarrow H_2$$

$$Fe_2O_3 + 6H^+ + 2e^- \longrightarrow 2Fe^{2+} + 3H_2O$$

At the anode, first magnesium anode is oxidized since it has a higher negative potential, and at the cathode, first $Fe_2O_3$ is reduced, since it requires a higher positive potential. After $Fe_2O_3$ layer is removed, hydrogen evolution becomes the only cathodic reaction leading to a negative change in the combined potential. Consecutively, after the magnesium anode is consumed, a positive change of about 50 mV occurs in the combined potential. However, even when the increase of 50 mV is considered, the combined potential remains more negative than the –780 mV potential of SCE, which is increased to that value from the initial value of –380 mV soon after the application of cathodic protection current.

### 16.1.3 Sacrificial Anode Cathodic Protection Current Need

For tanks with surfaces that have corrosion deposits and no coating, an average cathodic protection current of 40 $mA/m^2$ is initially needed, which is then reduced down to 15 $mA/m^2$ after about 2 months. Thus, in average, 20 $mA/m^2$ is deemed sufficient to provide cathodic protection for a period of 10 years until the next maintenance takes place, which can be provided by a certain quantity of aluminum anodes.

On the other hand, quantity and number of magnesium anodes depend primarily on the thickness of the rust layer that initially needs to be removed. Usually a current between 80 to 120 $A/m^2$ per hour is applied, while 50 $A/m^2$ per hour is considered the minimum current for removal of the oxide layer. A reference electrode, such as SCE, which is specially designed not to be damaged by the crude oil inside the tank, is installed in the tank to control the current supplied by the anodes and to control the tank-electrolyte potential.

Due to the salty nature of the corrosive liquid accumulated at the lower phase of the stored crude oil in the storage tank, all magnesium, zinc, and aluminum anodes can be used.

*i. Magnesium Anodes*

Potential difference between the magnesium anodes and the steel is about 650 to 700 mV, resulting in polarization of steel in a short period, which is why magnesium anodes can be used in high resistivity waters as well. However, magnesium anodes' current efficiency is around 50%. Their current capacities are also lower compared to other anodes; therefore, cathodic protection service life with magnesium anodes is considerably shorter, and as such they are not used for protection of crude oil tanks, since maintenance of crude oil tanks is performed every 5 to 10 years, which is a lot longer than the duration that magnesium anodes can provide cathodic protection.

### *ii. Zinc Anodes*

Current efficiency of zinc anodes is more than 90%; however, since the potential difference between zinc and steel is only 250 mV, it is difficult to withdraw currents of desired amounts. Thus, zinc anodes are commonly used at the beginning stages to polarize storage tank bases. As the steel polarizes, potential difference between the anode and the cathode is lessened, and thus the current withdrawn is automatically adjusted.

### *iii. Aluminum Anodes*

Indium alloyed aluminum anodes are the most appropriate anodes for cathodic protection of inner surfaces of crude oil storage tanks. Potential difference between the aluminum anodes and steel is also 250 mV; however, their current efficiency is 90% and they have 3.5 kg/A.year current capacities, translating into more economy, which allows them to be used for long durations, such as 5 to 10 years. 1 kg of indium alloyed aluminum anode can protect 1.5 $m^2$ surface area of the tank for 10 years.

### *iv. Mg + Al Dual Anodes*

When magnesium and aluminum dual anode systems are used, first, magnesium anodes that have more negative potentials act with their high potentials, polarizing the steel, and in the meantime, aluminum does not corrode, since they act as the cathode compared to the magnesium anodes. After the complete consumption of magnesium anodes, this time aluminum anodes become anodes and begin protecting the tanks from corrosion for long periods, given their high current capacities. It is possible to have a cathodic protection service life of up to 10 years, provided that respective masses of both types of anodes are calculated accurately.

Magnesium and aluminum dual anode systems are especially suitable for tanks with no coating, since magnesium anodes provide the high potential needed to remove the corrosion layer formed on the surface, which requires about 70 to 120 A.hour/$m^2$ current primarily for the reduction of iron (III) oxide film to iron (II) cations. As the oxide layer is removed, the required current is also reduced, and the system potential increases in the negative direction until the magnesium anodes are completely out of service. Afterwards, the system's potential begin to increase in the positive direction, and at that point, aluminum anodes begin operating, keeping the potential over the protection potential criterion of –0.850 V, providing the desired service life with their high current capacities.

### 16.1.4 Problems with Cathodic Protection of Storage Tanks

The biggest problem faced with implementation of cathodic protection is not being able to provide a uniformly distributed cathodic protection current at the tank base, resulting in a considerable potential difference, especially in very large tanks between the center of the tank and its periphery. As a result, difficulties occur, since potential of the center of the tank must be brought to the potential required by the protection criteria.

Additionally, it is not possible to accurately measure the potential of the center of the tank with reference electrodes, since they are installed outside of the tank. It is required to place the reference electrodes as much as near to the surface of which the potential will be measured; therefore, during the manufacturing and installation stages of the tanks, a reference electrode must be installed in a perforated tube underneath the center of the lower surface of the tank. Then, both the tank and the reference electrode are connected externally to the measurement unit.

For tanks that do not have a reference electrode installed underneath the center of the lower surface of the tank, it is not possible to measure the potential of the center of the tank; thus, the IR ohmic potential reduction from the periphery to the center of the tank must be calculated using the following formula:

$$\Delta E = \rho \frac{i \times r}{2} \tag{86}$$

where $\Delta E$ is the change in potential at a distance of r from the periphery of the tank, i is the cathodic protection current density in $A/cm^2$, and $\rho$ is the terrain resistivity in ohm.cm.

As the cathodic protection current density and terrain resistivity increase, the potential difference between the periphery and the center of the tank increases as well. As an example, for a tank that has a diameter of 72 m and for a cathodic protection current of $10^{-4}$ $mA/cm^2$ in a terrain of 5000 ohm.cm resistivity,

$$\Delta E = 5000(0.0001 \times 3600)/2 = 900 \text{ mV}$$

thus for –850 mV criterion to be satisfied at the center of the tank, potential must reach –850 mV + (–900 mV) = –1.750 V. However, in practice, potentials higher than 1.5 V based on CSE electrode are not wanted, to avoid overprotection. The following solutions are implemented to overcome this problem:

1. Anodes are placed around the tank's periphery equally apart from each other and 1.5 to 2 meters away from the tank's base and at least 1.5 meters vertically inside the ground. Anode cables are attached to the T/R unit's positive pole using a ring coil.

2. In the second method, anodes are placed underneath the tank horizontally and at equal distances from each other. However, this method can be applied only to new tanks.
3. In the third solution, uniform distribution of potential is achieved using an anodic bed that is 100 to 300 meters away from the tank. This solution would be economical if many tanks are cathodically protected, such as in a tank farm.
4. In another solution, if there is only one tank that needs to be cathodically protected, then anodes can be installed with an angle of 45 degrees to the tank's base, leading to a uniform current distribution at the tank's base, and also resulting in the potential at the center of the tank to rise up to the desired value. The depth of the anodic beds where the inclined anodes are placed can be between 7 to 10 meters, based on the size of the tank, so that the distance of the anode both to the center and to the periphery of the tank is at an optimum one.

# 17

# Corrosion and Cathodic Protection of Metallic Structures in Seawater

High conductivity of seawater makes it highly corrosive for metallic structures. The fact that chlorides prevent passivation is another factor increasing corrosion. Other major factors are temperature, pH, dissolved oxygen concentration, and fluidic rate. Additionally, whether the metallic structure is embedded in the seabed via poles or whether it is floating, e.g., ships or under the sea pipelines, or whether they are used for transportation, e.g., storage tanks, is also of utmost importance.

Salt level of the seawater is defined as the dissolved salt amount in 1 kg of seawater. In natural seawater, salt levels are proportional with the chloride concentration based on the following formula, which is stated in equation 4 discussed in section 4.1.2, "Corrosion in Seawater and in Fresh Waters":

$$\textit{salt level}\,(g/kg) = 1.80655 \times \textit{chloride concentration}\,(g/kg)$$

Salt level of the sea also increases with increasing sea depth, with an increase of about 0.5 g/kg for every 100 meters of depth.

## 17.1 Factors Affecting Corrosion Rate of Metallic Structures in Seawater

There are many factors affecting the rate of corrosion reactions taking place in seawater such as conductivity or resistivity of the water, pH, temperature, dissolved oxygen concentration, fluidic rates, etc.:

### 17.1.1 Effect of Resistivity on Corrosion in Seawater

The most important aspect of seawater in terms of corrosion is its resistivity, since it is low, around 16 to 40 ohm.cm, leading to severe corrosion to take place. Resistivity is the opposite of conductivity (ohm/cm), and thus their relation is inversely proportional. As the salt content increases, conductivity increases, and thus conductivity of seawater is about 250 times more than fresh water. Rising temperatures also increase the conductivity.

Resistivity of seawater can also be found based on the density, which is usually the preferred method, since it is easier to determine the density than to determine the salt concentration. However, since density changes with temperature, measurement temperature must also be noted.

### 17.1.2 Effect of pH on Corrosion in Seawater

Regularly, pH of the sea is between 8.1 and 8.3 due to the equilibrium between the bicarbonate ions present in the seawater and the carbon dioxide gas present in the atmosphere. Carbon dioxide that is dissolved in the seawater is used for photosynthesis by the plants living in the sea. Since sunlight is abundant close to the sea surface, photosynthesis occurs more commonly at the sea surface, consuming the carbon dioxide, and thus resulting in an increase in pH. On the other hand, a reduction of pH occurs as the sea depth increases due to the carbon dioxide and hydrogen sulfide that are produced by the rotten organic materials. Anaerobic decomposition processes produce hydrogen sulfide of up to 50 ppm, making the environment very suitable for microbial corrosion. Thus, pH levels of 8.2 at the surface of the sea reduce to pH 7.6 at around 100 meters below the surface.

Due to high pH levels, cathodic reaction of corrosion process is the oxygen reduction reaction, and above ph 8, corrosion products precipitate at the metal surfaces in the form of carbonates and hydroxides, with cations present in the seawater such as calcium and magnesium cations leading to the formation of a shell over the metal surface, preventing further corrosion from taking place. The chemical and physical composition of this shell also depends on the cathodic protection current intensity. At high current intensities, the shell is formed faster and includes more magnesium, resulting in a softer, spongy shell. Formation of the shell prevents corrosion by slowing down the diffusion of oxygen to the metal surface and by increasing the electrical resistivity. Therefore, current intensity needed for impressed current cathodic protection decreases to one-third after a few months.

### 17.1.3 Effect of Temperature on Corrosion in Seawater

Regularly, increasing temperatures increase corrosion rate as well. However, increasing temperatures reduce the solubility of oxygen in seawater. Furthermore, high seawater temperatures also provide a suitable habitat for living organisms, thus increasing their quantities. These living organisms produce a shell over the metal surface due to fouling effect, which reduces the diffusion of oxygen to the metal surface, similar to the shell formed by the carbonates and hydroxides. Consequently, the corrosion rate increases with temperature up to a certain limit, and decreases thereafter.

Based the seasons and on the geographical regions, the temperature of the sea varies, usually between 5°C and 25°C. However, seasonal changes are effective only down to a certain depth. Beyond 50 meters of depth, meteorological events are no longer effective, and temperature of the seawater remains constant at 4°C to 5°C.

### 17.1.4 Effect of Dissolved Oxygen Concentration

Dissolved oxygen concentrations in the seawater decrease with increasing temperatures, and also with increasing salt contents. 7 ml/L solubility at 5°C decreases to 4.7 ml/L at 25°C; while the same solubility values in distilled water are 8.9 and 5.8 ml/L, respectively. Dissolved oxygen concentration also decreases with increasing depth, which becomes half of what it is at the surface at about a 300 meter depth. As a reason, dissolved oxygen concentration at areas close to the sea surface is at equilibrium with the atmospheric oxygen; therefore, it is the highest at the sea surface. Diffusion of oxygen in water is very slow to the depths of the sea. Another factor resulting in relatively higher dissolved oxygen concentrations at the sea surface is the production of oxygen via photosynthesis realized by living organisms that live at levels close to the sea surface due to presence of abundant sunlight and carbon dioxide.

Rotten organic materials, as well as nitrates and phosphates, reduce the amount of oxygen since it is consumed by associated chemical processes; however, this does not reduce the corrosion rate, since, in contrast with the impeded corrosion processed caused by the reduction in the amount of oxygen, a very suitable environment for anaerobic microbial corrosion is established.

The fact that corrosion rate mainly depends on the dissolved oxygen concentration reveals itself in structures that are embedded in the seabed, since corrosion rate for such metallic structures is higher at sea surface and lower at the seabed. This type of corrosion that is specifically based on oxygen concentration is named "Waterline Corrosion" and is discussed in section 4.2.16.

Therefore, for metallic structures that are in seawater environments, the following cases occur:

- Above the sea surface, where metallic structures cannot be wetted by the seawater, there is a very high corrosion rate due to a humid atmosphere having high salt contents.
- At the sea surface, which is also specifically called the "splash zone," metal surfaces get in cycles of wet and dry, getting wet and then consecutively drying up; thus, the highest corrosion rate occurs at this level, since all reactants of corrosion reactions are abundantly present, and also the protective shells cannot form since it is continuously being washed.
- At the level just below the sea surface, the corrosion rate is a lot lower, and only second to the seabed, since surfaces of the metallic structure remain either dry or wet based on the time of the tides for long periods of time, not getting into cycles of wet and dry, which allows the formation of protective shells over the metal surfaces, which is also the reason why corrosion rate of steel structures in the sea is the highest within the first month, while it decreases gradually thereafter.
- The level that is just below the tide region always remains inside the sea, and thus it is wet, and it also has a very high corrosion rate, primarily since there is very high dissolved oxygen concentrations at this level due to being close to the sea surface, while deeper than 2–3 meters below the sea surface, oxygen concentration decreases, and the corrosion rate remains constant thereafter.
- The last region is the region where the metallic structure is embedded in the seabed. There, the metallic structure is covered by the seabed mud, and thus that part of the metallic structure normally has the lowest corrosion rate, unless there is high anaerobic corrosion taking place.

### 17.1.5 Effect of Fluid Rate

Fluid rate affects corrosion, since a fast flowing fluid increases the amount of oxygen reaching the metal surface, preventing the formation of a protective shell. In stagnant conditions, corrosion rate is in general 0.013 mm/year, which increases to 0.050 mm/year for fluids moving with a fluidic rate of 1.5 mm/second, to 0.074 mm/year for fluids moving with a rate of 3.0 mm/second, to 0.084 mm/year for fluids moving with a rate of 4.5 mm/second, and to 0.089 mm/year for fluids moving with a rate of 5.0 mm/second.

This situation reveals itself during the cathodic protection of ships, since the current intensity needed for cathodic protection increases twofold while the ship is in motion.

## 17.2 Cathodic Protection of Metallic Structures in the Sea

Due to high corrosivity of the seawater, it is not possible to prevent corrosion of metallic structures in the sea by only applying paint coatings. Although in theory, metallic structures can be repainted, as in the example of ships, this is not possible when considering the metallic structures that are fixed in the seabed. Thus, the most effective corrosion prevention method that stands out is cathodic protection. Since resistivity of the sea water is low, both sacrificial anode and impressed current cathodic protection systems are suitable.

### 17.2.1 Cathodic Protection Current Need

For steels dipped in stagnant seawaters, cathodic protection current need is around 150 $mA/m^2$, which rises to 300 $mA/m^2$ and up to 350 $mA/m^2$ in wavy seas and at levels close to the sea surface. However, due to polarization and formation of protective shells, after a while, the cathodic protection current intensity is reduced down to one-third of the initially required magnitudes.

Cathodic protection current needs vary up to tenfold between different ships depending upon the type of the ship, quality of the paint, loads and speeds of the ships, and upon the salt level and the temperature of the sea that the ships are cruising or sailing within. The total current needed for cathodic protection of ship over a year also depends on the time that the ship is anchored at a port, which changes the cathodic protection current need from 10 $mA/m^2$ up to 60 $mA/m^2$. Specifically, current intensity needed for cathodic protection of propellers and the wheel can be around 500 $mA/m^2$, since they are bare metallic components, and presence of turbulence effects around these compounds results in maximum access of oxygen. Further, turbulence also results in cavitation corrosion. Furthermore, propellers and wheels made of bronze are susceptible to galvanic corrosion, which also causes an increase in the current needed for cathodic protection.

### 17.2.2 Cathodic Protection Criteria

General –850 mV protection criteria is appropriate for protection of metallic structures in the seawater, especially since the resistivity of the sea water is low, and thus ohmic potential reduction can be omitted. However, use of copper/copper sulfate electrode as the reference electrode is not appropriate due to chloride contamination. Instead, silver/silver chloride electrode or pure zinc electrodes are used. Steel structures' protection potential in seawater compared to silver/silver chloride reference electrode is –0.760 V; however, if the reference electrode is used directly, then the protection potential must be taken as –0.800 V. Pure zinc's potential in the seawater does not change with variations in salt content or flow rates. Zinc electrode's potential in the seawater compared to saturated copper/copper sulfate electrode is –1.100 V, and thus the protection potential occurs as 0.25 V.

## 17.3 Cathodic Protection of Ships

Due to the aggressive nature of the seawater and also due to fouling effect caused by the organisms that live close to the sea surface, even the most durable paints can last only a few months; thus, cathodic protection is administered along with painting of the ships for maximum corrosion protection, and while doing that, corrosion rates are not estimated beforehand, since there is a tendency to employ both measures anyway. Consequently, corrosion of ships depends on different factors:

- Cleanliness of the metal surface before the paint is applied;
- Type of the paint used on the body of the ship;
- Thickness of the paint and its durability;
- Type of the ship, its speed and weight;
- The durations the ship is anchored at a port and durations that it is in motion; and
- Salt content and the temperature of the seawater ship is cruising or sailing.

Since it is difficult to determine all of these factors, ships are protected from corrosion usually by both application of paint and cathodic protection without making estimation about the corrosion rate.

### 17.3.1 Painting the Ships

Different paints must be used when painting a ship, since there are differences between different parts of the metallic body of the ship in terms of the level of being exposed to the seawater. The regions of the ship that are in contact with seawater must not be painted with paints that can saponify, since sea water has an alkaline character. Further, the top coat of the paint must contain poisons to prevent a fouling effect. Additionally, quickly drying paints should be used so that the waiting time of the ship at the dry pool is lessened.

Another use of the painting application is to reduce the frictional forces due to the presence of the corrosion products, shell formation, and living organisms, leading to millions of dollars of savings.

### 17.3.2 Sacrificial Anode Cathodic Protection of Ships

Ships were initially protected only via sacrificial anode cathodic protection systems that employ magnesium anodes, but due to the low current efficiencies and capacities of magnesium anodes, sacrificial anode cathodic protection systems can only be used for short durations. Another problem faced with the use of magnesium anodes is that their potentials are high, leading to overprotection, and thus damaging the paint coating. Hence, zinc and aluminum anodes are preferred over magnesium anodes. Potential difference between the zinc and aluminum anodes and with that of the steel structure to be protected in seawater is 0.250 V, which decreases with increasing cathode potential. Potential of the ship can rise up to 0.95 V at most, compared to Ag/AgCl electrode. As the ship's potential reaches the optimum value needed for cathodic protection, current withdrawn from galvanic anodes begins to decrease. On the contrary, if the ship's potential is low, then the potential difference between the anode and the cathode increases, and thus the amount of current withdrawn from the anodes increase as well. As such, the current withdrawn from the anodes is automatically adjusted.

*i. Number of Anodes and Anodic Mass*

Calculation of the total current needed for cathodic protection of a ship is done when the ship is fully loaded, so that total area of surfaces that get wet can be accurately calculated beforehand. Currents needed for cathodic protection of the wheel and the propeller are also separately added. Cathodic protection service life is determined based on the ship's re-entry time to the dry pool, which usually takes around 1.5 to 2 years. Thus, the anodic mass needed for cathodic protection, during this service life of 1.5 to 2 years, is calculated as follows:

$$\text{Anodic mass} = \frac{\text{anode life(years)} \times \text{current withdrawn(A)} \times \text{anodic capacity(kg/A.year)}}{\text{anodic efficiency} \times \text{usage factor}} \quad (87)$$

Usage factor depends on the shape of the anode; for instance, if the anode is of cylindrical shape, the usage factor is 0.90, if it is rectangular, then the usage factor is 0.85, and if it is trapezoid, then the usage factor is 0.80. Consequently, the number of anodes needed for cathodic protection can be found by dividing the total anodic mass needed for cathodic protection with that of the mass of one anode given its current efficiency, capacity, and usage factor, along with the cathodic protection current need.

### *ii. Resistances of Anodes*

Resistances of anodes, which are installed on bases that are more than 30 cm high and used for cathodic protection of metallic structures stationed in sea, are calculated as follows, given that they have more than 10 times lengths of their effective radii:

$$\mathrm{R} = \frac{\rho}{2\pi L}\left(\ln\frac{2L}{r} - 1\right) \tag{88}$$

where $\rho$ is the resistivity of seawater in ohm.cm, L is the length of the anode in cm, and r is the effective radius in cm. In anodes of rectangular shape, effective radius is calculated as follows:

$$r = \sqrt{\frac{a}{\pi}} \tag{89}$$

where a is the anode's cross-sectional area. Cross-sectional area of the anode is reduced as it is used, and calculation of effective radius to determine the anodic resistance is revised, assuming that the cross-sectional area of the anode can be reduced down to 60% of the original area:

$$r = \sqrt{\frac{a \times 0.60}{\pi}} \tag{90}$$

Resistances of anodes that have trapezoid shape are calculated using the following formula:

$$R = \frac{\rho}{a + b + c} \tag{91}$$

where $\rho$ is the resistivity of seawater in ohm.cm and a, b, and c are the dimensions of the anode.

*iii. Anodic Current Capacity*

After the resistances of anodes are calculated, the maximum current that can be withdrawn from a single anode is calculated using the following formula:

$$i = \frac{\Delta E}{R} \tag{92}$$

where $\Delta E$ is taken as 0.250 V in the case of zinc and aluminum anodes. Consequently, in addition to the calculation of the minimum number of anodes needed for cathodic protection using the total anodic mass from formula 84 and the mass of a single anode, it can also be calculated via dividing the total current needed for cathodic protection with the maximum current that can be withdrawn from a single anode:

$$\#\ of\ galvanic\ anodes\ needed = i_{total} / i_{\mathrm{single}\ anode} \tag{93}$$

*iv. Anode Distribution*

Anodes are usually welded onto the two sides of the ship symmetrically and about 1 meter under the water level. The side of the anode that is in contact with the ship's body is painted with a thick coating. It is not possible to protect a large area of the ship with a single anode due to the low resistivity of the seawater, which results in a reduction in the potential of the current withdrawn from the anode, even in short distances. Thus, usually only one anode is installed for a wet surface area of 10 $m^2$, and another is installed at every 7 or 8 meter distances, which are in addition to the anodes that are used for the wheel and the propeller.

### 17.3.3 Impressed Current Cathodic Protection of Ships

Impressed current cathodic protection is economically more feasible for cathodic protection of relatively bigger ships. Additionally, it allows automatic and simultaneous adjustment of the current, as the cathodic protection current need frequently varies due to factors such as:

- Total wet surface area of the ships varies substantially, up to 50% based on their loads.
- Cathodic protection current need is the lowest when ships are sitting at the dock, which increases when they are in motion.
- Ships with bad paint qualities in particular may require 2 to 3 times more cathodic protection current.
- Cathodic protection current need increases with increasing temperatures as well. For instance, current needed for cathodic protection is 20% to 25% more in tropical waters than in cold waters.

Most commonly, silver/lead and platinum coated titanium anodes are used for impressed current cathodic protection of ships. These anodes are installed in a special way on the ship's body so that frictional losses are reduced as much as possible. Additionally, the surrounding areas where the anodes are installed must be insulated using a thick paint coating or fiberglass to prevent the damages caused by the high potentials applied from the anodes. The radius of the area that must be insulated is calculated via the following formula:

$$\Delta E = \frac{\rho I}{2\pi r} \tag{94}$$

where $\Delta E$ is the potential change at a distance of r (cm), $\rho$ is the resistivity of seawater in ohm.cm, and i is the current intensity withdrawn from the anode in Amperes.

It must be ensured that each and every point on the ship's body has a potential of more than –0.760 V in the negative direction compared to silver/silver chloride electrode for the cathodic protection to occur. On the other hand, if the potentials are kept too high, then the coating may be damaged as well, and thus difference in potentials ($\Delta E$) should not be over than 1.0 V.

### 17.3.4 General Guidelines about Cathodic Protection of the Ships

1. Cathodic protection of ships begins even before the ships start to be used.
2. Cathodic protection duration is usually calculated based on the next renovation stage, which is usually in about 1.5 to 2 years.
3. Both sacrificial anode and impressed current cathodic protection methods are applicable to ships. Choice of cathodic protection is based on economical factors; however, commonly, sacrificial anode cathodic protection is administered to small ships, and impressed current cathodic protection is administered to big ships.
4. In both cathodic protection systems, anodes are attached to the ship's body; however, these attached anodes negatively affect the ship's speed, due to frictional forces they cause, leading to more oil consumption, which is estimated around 1% of the total consumption. Thus, anodes of different shapes such as of trapezoid shape are developed to reduce such frictional forces.
5. Although ships' propellers are in contact with the body of the ship, due to presence of oils, electrical resistances are formed in between the propeller and the ship. Since the resistance must be reduced below 0.0001 ohm for cathodic protection, a bronze brush is attached, forming an electrical bond between the two parts.
6. More often in impressed current cathodic protection systems, high potentials occur at areas close to where the anodes are installed, damaging the paint layer.
7. Cathodic protection current needs of ships vary over time, primarily due to paint that is wearing out.

## 17.4 Cathodic Protection of Pier Poles with Galvanic Anodes

The best method to prevent corrosion of pier poles in seawater is cathodic protection; however, it is not possible to cathodically protect the splash zone regions; thus, these regions are usually coated with cement or another appropriate material for corrosion prevention.

It is very important to determine the right amount of cathodic protection current, since corrosion rate is very different at different levels of the pole. Thus, when surfaces of bare steel poles are considered:

- In wavy and rough seas, at first, the cathodic protection current need is 350 $mA/m^2$, and later, after polarization takes place, it is reduced down to 100 $mA/m^2$.
- In stagnant seawaters, at first, the cathodic protection current need is 250 $mA/m^2$, and later, after polarization takes place, it is reduced down to 70 $mA/m^2$.
- At seabed in mud, at first, the cathodic protection current need is 50 $mA/m^2$, and later, after polarization takes place, it is reduced down to 15 $mA/m^2$.

In the case of coated steel pile surfaces:

- In wavy and rough seas, at first, the cathodic protection current need is 50 $mA/m^2$, and later, after polarization takes place, it is reduced down to 15 $mA/m^2$.
- In stagnant seawaters, at first, the cathodic protection current need is 30 $mA/m^2$, and later, after polarization takes place, it is reduced down to 10 $mA/m^2$.

- At seabed in mud, at first, the cathodic protection current need is 10 $mA/m^2$, and later, after polarization takes place, it is reduced down to 5 $mA/m^2$.

It is also important to note that coating of the parts of the pile that lay at lower levels of the sea can be damaged up to 40 to 50% during the installation of the metallic structure in the seabed, while at other regions, an estimated damage of 5 to 10% damage is considered normal.

# 18

# Cathodic Protection of the Potable Water Tanks

Resistivities of potable waters are very high, and thus inner surfaces of the potable water tanks made of steel can be protected from corrosion for long durations by painting them with a suitable paint. However, if chlorine is used as disinfectant, pitting corrosion occurs, necessitating the application of cathodic protection along with paint applications.

Two things need attention in cathodic protection of potable water tanks; first is not using anodes that can release harmful ions into the potable water, and second, tanks must not be entered for reasons such as measurement, maintenance, or repairs. Controls must be performed externally, and service life of cathodic protection must be adjusted as long as possible. Since potable waters have high resistivities, in theory, impressed current cathodic protection systems seem more reasonable; however, the current needed for cathodic protection of a well-coated water tank is very low, and thus sacrificial anode cathodic protection is more economical, along with other advantages cited below:

- There is no need for installation of an additional cable system to provide electricity.
- There is no need for maintenance and adjustments.
- Overprotection and related peeling off of the coating does not occur.
- Magnesium cations produced by the dissolution of magnesium anodes do not cause any problems; thus, HP magnesium anodes can be used in waters up to 10000 ohm.cm.

# 19

# Corrosion and Corrosion Prevention in Boilers

Boilers are equipment where fuels are converted to heat to obtain steam out of water. Corrosion types encountered in boilers in water treatment technology are oxygen corrosion, carbon dioxide corrosion, caustic corrosion, acidic corrosion, hydrogen embrittlement, and crevice corrosion. The primary reasons for corrosion in boiler tank systems are low pH and dissolved corrosive gases such as oxygen and carbon dioxide. Low pH corrosion results in thinning of the boiler pipes, while dissolved oxygen leads to pitting corrosion. Another corrosion type is caustic corrosion leading to caustic cracking. Nowadays, caustic corrosion is rarely encountered in modern tanks, since now they are all welded. In hydrogen embrittlement, hydrogen originated from corrosion reactions diffuse into metal at high temperatures, reacting with the carbon in boiler steel, resulting in decarburization due to formation of methane, which leads to an increase in pressure, resulting in intergranular corrosion and cracks.

## 19.1 Corrosion in Boilers

Corrosion in boiler tanks is generally due to gases, alkali bases, acids, and galvanic effects. Consequently, dissolution of the boiler metal is directly proportional with dissolved oxygen and carbon dioxide concentrations, temperature, and flow rate of water, while it is inversely proportional with the alkalinity of the environment, hardness of the water, and pH. Galvanic corrosion occurs when different metals or alloys come into contact, while acidic and basic corrosion are due to the increasing conductivity of the electrolytes; thus, pH has to be kept over a minimum level for the formation of corrosion-preventative calcium phosphate precipitates. On the other hand, if the pH is too high, caustic corrosion may occur. The major corrosion processes that occur in boiler systems are summarized as follows:

*i. Acidic Corrosion*

First, HCl is formed, reacting with iron afterwards, leading to its dissolution from the bulk into the solution phase in the form of chlorides:

$$MgCl_2 + 2H_2O \xrightarrow{\Delta} Mg(OH)_2 + 2HCl \qquad \text{(Eq. 130)}$$

$$Fe + 2HCl \longrightarrow FeCl_2 + H_2 \qquad \text{(Eq. 131)}$$

*ii. Caustic (Basic) Corrosion*

Soft water is used in boilers. For that reason, sodium carbonate is used in the softening (lime soda) process. Failure of boiler may occur due to lime soda process when $Na_2CO_3$ dissociates into caustic soda and carbon dioxide at high pressures according to the following reaction:

$$Na_2CO_3 \longrightarrow 2NaOH + CO_2 \qquad \text{(Eq. 132)}$$

Hence, concentration of caustic soda increases and ultimately attacks the boiler to dissolve out iron as sodium ferrate, as shown in reactions 133 through 136. This causes embrittlement of boiler walls, leading to failure of the boiler.

$$2Fe + 2NaOH + 2H_2O \longrightarrow 2NaFeO_2 + 3H_2 \quad \text{(Eq. 133)}$$
$$Fe_3O_4 \longrightarrow Fe_2O_3 + FeO \quad \text{(Eq. 134)}$$
$$Fe_2O_3 + 2NaOH \longrightarrow 2NaFeO_2 + H_2O \quad \text{(Eq. 135)}$$
$$FeO + 2NaOH \longrightarrow NaFeO_2 + H_2O \quad \text{(Eq. 136)}$$

*iii. Gas Corrosion*

Bicarbonates and carbonates in the feeding water decompose at high temperatures and pressures produce carbonic acid in the steam-condensation units. Produced carbonic acids are neutralized with ammines, of which some have high molecular weights, and thus are able to form a protective layer on the surface of the metal. Further, when the system is out of service, oxygen corrosion and sometimes low pH corrosion become important. The following reactions summarize the major corrosion reactions due to gases taking place in boilers:

$$2HCO_3^- \xrightarrow{\Delta} CO_3^{-2} + CO_2 + H_2O \quad \text{(Eq. 137)}$$
$$CO_3^{-2} + H_2O \xrightarrow{\Delta} 2OH^- + CO_2 \quad \text{(Eq. 138)}$$

Carbon dioxide, formed by decomposition of bicarbonates and of carbonates, reacts with water, yielding carbonic acid:

$$CO_2 + H_2O \longrightarrow H_2CO_3 \quad \text{(Eq. 139)}$$

Produced carbonic acid reacts with the iron, forming iron bicarbonate in reaction 140:

$$Fe + 2H_2CO_3 \longrightarrow Fe(HCO_3)_2 + H^+ \quad \text{(Eq. 140)}$$

Iron bicarbonates can react with dissolved oxygen, forming oxides, as seen in reaction 84, while in the absence of the dissolved oxygen, they can also react with water, producing hydroxides, as seen in reaction 85:

$$4Fe(HCO_3)_2 + O_2 \longrightarrow 2Fe_2O_3 + 8CO_2 + 4H_2O \quad \text{(Eq. 141)}$$
$$FeHCO_3 + 2H_2O \longrightarrow Fe(OH)_3 + H_2CO_3 \quad \text{(Eq. 142)}$$

Corrosion reactions of the boiler metal due to dissolved oxygen first yield iron (II) cations, as shown in reaction 143, and then iron (III) cations, as shown in reaction 144.

$$2Fe + O_2 + 4H^+ \longrightarrow 2Fe^{2+} + 2H_2O \qquad \text{(Eq. 143)}$$

$$4Fe^{2+} + O_2 + 4H^+ \longrightarrow 4Fe^{3+} + 2H_2O \qquad \text{(Eq. 144)}$$

### *iv. Corrosion due to Boiler Water Hardness*

Boilers are exposed to corrosion as well as deposits because of the hardness of water. Hardness in water is either temporary hardness or permanent hardness, while total hardness includes both. Temporary (alkaline) hardness is originated from the calcium and magnesium bicarbonates present in water. When water is heated, chemicals that cause temporary hardness decompose, leading to the formation of carbon dioxide, calcium carbonate, and magnesium hydroxide:

$$Ca(HCO_3)_2 \longrightarrow CaCO_3 + H_2O + CO_2 \qquad \text{(Eq. 145)}$$

$$Mg(HCO_3)_2 \longrightarrow Mg(OH)_2 + 2CO_2 \qquad \text{(Eq. 146)}$$

Compounds that cause permanent hardness are magnesium and calcium sulfates, silicates, chlorides, nitrates, and various iron oxide compounds. These compounds are neutral, and they do not decompose with heat. They cause hard and soft deposits on surfaces that transfer heat. Hard deposits are scales and soft deposits are sludge. Both deposits lead to crevice corrosion, heat loss, and increase in temperature of boiler pipes. Eventually, pipes of the boiler tank system get blocked due to lessened circulation of boiler system water.

HCl and $H_2SO_4$ are used to clean the deposits, and inhibitors are used to prevent corrosion of steel equipment after getting cleaned with strong acids.

### *v. Corrosion due to Impurities in Boiler Water*

Another problem in boiler systems is carryover that is caused by chemicals that are dissolved or insoluble in boiler water, which are carried to vapor phase via water vapor. These particles that are carried over form deposits at locations such as valves and condensation conversion systems. Boiler tank systems, those having pressures over 40 bar, cause silicates to become volatile, resulting in carryover occurring, leading to formation of deposits at turbines.

## 19.2 Corrosion Prevention in Boilers

Prevention of corrosion in boilers is important since components of boilers, which are economizers, steamers, and heaters, are all made of steel alloys and are susceptible to corrosion. Corrosion prevention is primarily done by increasing the pH of the boiler feeding water, removal of dissolved gases such as oxygen and carbon dioxide, passivation of metal surface with a magnetic layer, limiting and monitoring the free hydroxide, silica, and chloride ion concentrations, and lastly, via preventing the deposit formation by employing periodic cleaning procedures. For that reason, boiler waters must be cleaned at the source and before they are in the system. Methods such as filtration, use of cation resins, demineralization, and reverse osmosis are among common employed techniques to clean boiler waters. Secondly, protective chemicals such as phosphates must be used when water is in the boiler tank system. Phosphates form muddy compounds with calcium salts in alkaline media, which can be removed later via bluffing, which is the removal of contaminated portion of the water. Other protective chemicals preventing corrosion of boiler systems include organic polymers or chelating agents such as EDTA, oxygen scavengers, and amines to neutralize carbon dioxide. Removal of dissolved gases is commonly performed either via degassing or via chemical conditioning, such as use of scavengers.

### 19.2.1 Degassing

Gas corrosion that is due to the dissolved corrosive gases in the boiler feeding water is prevented with degassers and by chemical conditioning. However, it is important to note that even if degassing process is very efficient, 15% to 20% of the oxygen still remains in the system. Additionally, if the recommended temperature range of 102°C–105°C cannot be reached during the process, oxygen may be dissolved in the feeding water even more.

### 19.2.2 Chemical Conditioning

Chemical conditioning is done via chemicals such as sodium sulfite, hydrazine, morpholine, volatile ammines, etc. Volatile ammines are oxygen scavengers that form protective layers, absorb corrosive gases, and also neutralize acids. Another chemical conditioner, hydrazine, is carcinogenic, and also causes corrosion of copper and copper alloys. On the other hand, reactions of sodium sulfite are very slow at low temperatures; thus, copper salts are added to the mix to accelerate its reactions. Additionally, due to formation of corrosive $SO_2$ and $H_2S$ gases, the maximum allowed sodium sulfite use as a chemical conditioner is 60 $kg/cm^2$. Among the sulfites used are sodium sulfite, sodium bisulfate, and sodium metabisulfite; while the first is basic, with pH values of 8 to 9, the remaining two are weakly acidic. Common oxygen scavengers are:

*i. Sodium Sulfite ($Na_2SO_3$)*

Sodium sulfite reacts promptly with oxygen; however, since the produced sulfate increases the conductivity of the boiler water, process must be performed carefully:

$$Na_2SO_3 + \tfrac{1}{2} O_2 \longrightarrow Na_2SO_4 \quad \text{(Eq. 147)}$$

$$4\,Na_2SO_3 + 2\,H_2O \longrightarrow 3Na_2SO_4 + 2NaOH + H_2S \quad \text{(Eq. 148)}$$

$$Na_2SO_3 + 2\,H_2O \longrightarrow 2NaOH + H_2SO_3\,2H_2O + SO_3 \quad \text{(Eq. 149)}$$

$$Na_2SO_3 + H_2O \longrightarrow SO_2 + 2NaOH \quad \text{(Eq. 150)}$$

*ii. Hydrazine ($N_2H_4$)*

Hydrazine is commonly used as a oxygen scavenger in high pressurized boilers since it is a very effective scavenger, and its reactions with oxygen do not lead to an increase in the salt amount of the water, and thus do not increase the conductivity of the water. However, it is carcinogenic.

$$N_2H_4 + O_2 \longrightarrow 2H_2O + N_2 \quad \text{(Eq. 151)}$$
$$N_2H_4 + O_2 \longrightarrow N_2 + 2H_2O \quad \text{(Eq. 152)}$$

Hydrazine also assists the passivation process of the metal:

$$N_2H_4 + 6Fe_2O_3 \longrightarrow 4Fe_3O_4 + 2H_2O + N_2 \quad \text{(Eq. 153)}$$

On the other hand, hydrazine causes corrosion of copper and its alloys reacting with copper (II) oxide formed via oxidation of elemental copper and copper (I) oxide:

$$Cu + \tfrac{1}{2}O_2 \longrightarrow CuO \quad \text{(Eq. 154)}$$
$$2Cu_2O + O_2 \longrightarrow 4CuO \quad \text{(Eq. 155)}$$
$$4CuO + N_2H \longrightarrow 2\,Cu_2O + 2H_2O + N_2 \quad \text{(Eq. 156)}$$

Copper (II) oxide also reacts with ammonia and ammonium hydroxide forming complexes:

$$CuO + 4NH_3 + H_2O \longrightarrow Cu(NH_3)_4^{2+} \quad \text{(Eq. 157)}$$
$$CuO + 4NH_4OH \longrightarrow Cu(NH_3)_4(OH)_2 + 3H_2O \quad \text{(Eq. 158)}$$

*iii. Diethylhydroxylammine (DEHA)*

Use of diethylhydroxylammine (DEHA) has become more common, since use of sodium sulfite as an oxygen scavenger increases the conductivity of the solution, while hydrazine is carcinogenic. Diethylhydroxylammine (DEHA) is a volatile compound with corrosion protective effects in steam-condensation units. For every 1 ppm of oxygen, 3 ppm of

DEHA is used. Over temperatures of 288°C, diethylhydroxylammine (DEHA) decomposes into ammonia.

### *iv. Hydroquinones*

Hydroquinones have the highest reaction speed, and are also passivators. They do not decompose up to 304°C, and thus do not lead to an increase in the conductivity of the solution. For every 1 ppm of oxygen, 12 ppm of hydroquinone is used. Hydroquinones are volatile compounds, as is diethylhydroxylammine (DEHA); thus, they have corrosion protective effects in steam-condensation units.

### *v. Carbohydrazides*

Carbohydrazides are derivatives of hydrazine; thus, they are converted into hydrazine in boiler conditions as a result of removing oxygen. Use of carbohydrazides prevents its handlers from being exposed to the hydrazine's carcinogenic effect. It should be used with catalysts, due to its slowly proceeding reaction. For every 1 ppm of oxygen, 1.4 ppm of carbohydrazides is used.

### *vi. Isoascorbic Acid*

Isoascorbic acid (vitamin C) is used in chemical conditioning of boiler water along with ammonia or ammines. It has a good reaction speed, and for every 1 ppm of oxygen, 11 ppm isoascorbic acid must be used. Isoascorbic acid is not volatile; thus, it does not have any protective effect in steam-condensation units.

### *vii. Methylethylketoxine*

Methylethylketoxine can react with $O_2$ at every pH level. It acts as a passivator as well. Additionally, methylethylketoxine is volatile; thus, it prevents corrosion at the steam-condensation process line. For every 1 ppm oxygen, 5.4 ppm of methylethylketoxine is needed.

# 20

# Corrosion and Corrosion Prevention in Geothermal Systems

## 20.1 Corrosion in Geothermal Systems

Systems using geothermal fluids for heating and other purposes are also susceptible to corrosion depending on the chemical composition of the fluids. Geothermal fluids commonly have dissolved oxygen and ions such as $H^+$, $Cl^-$, $H_2S$, $CO_2$, $NH_3$, $SO_4^{2-}$, $Na^+$, $HCO_3^-$, $CO_3^{2-}$, $Ca^{2+}$, and $Mg^{2+}$, which all contribute to conductivity, and thus lead to corrosion. $Cl^-$, for instance, causes pitting corrosion, while all $Cl^-$, $SO_4^{2-}$, $Ca^{2+}$, and $Mg^{2+}$ salts cause permanent hardness, leading to deposits, resulting in problems in geothermal systems, as it is the case in oil production equipment.

High pH values increase carbonate concentrations, and thus increase calcium carbonate deposits. On the other hand, dissolved $CO_2$ leads to carbonic acid formation, which lowers pH and decreases calcium carbonate deposits; thus, adjusting the pressure to prevent $CO_2$ evolution limits calcium carbonate deposition.

At low temperatures, and for $Cl^-$ concentrations close to 100 ppm, corrosion is directly proportional with the square root of $Cl^-$ concentration, while over 50°C, $Cl^-$ concentrations of around 5 to 10 ppm are sufficient to initiate stress corrosion cracking for stainless steels. Sulfate reducing bacteria is another primary inducer of corrosion in geothermal systems. Sulfate reducing bacteria operate via the following reactions:

$$2CH_2O + SO_4^{2-} \longrightarrow H_2S + 2HCO_3^- \quad \text{(Eq. 159)}$$

$$Me^{2+} + H_2S \longrightarrow MeS + 2H^+,\ Me = \text{Metal} \quad \text{(Eq. 160)}$$

## 20.2 Corrosion Prevention in Geothermal Systems

In prevention of corrosion of underground steel pipelines carrying geothermal heating water, cathodic protection is widely implemented, provided that the following criteria are taken into account:

1. In sacrificial anode cathodic protection systems, cost of the applied current is higher than it is in the impressed current cathodic protection systems. Thus, especially in geothermal systems where high currents are needed, impressed current cathodic protection systems are more economical.
2. Additionally, it may not be possible to withdraw sufficient amounts of current from galvanic anodes at grounds with high resistivities if galvanic anodes

that have very low potentials compared to steel are used. Hence, among the metallic components in the system, the ones that have low potentials, possibly due to the differences in their coatings, are first associated with the galvanic anodes, increasing their low potentials above a certain limit so they can remain protected from corrosion. This way, the potential difference between the different metallic structures in the system becomes very small, and thus even if the impressed current cathodic protection system goes out of service, no interference occurs between underground steel pipelines carrying geothermal heating water.

# References


Abbot, D. (1966) *Essential of Sixth Form Inorganic Chemistry*, J. M. Dent and Sons Ltd., London.

Abboud, Y.; Abourruche, A.; Saffaj, T.; Berrada, M.; Charrouf, M.; Bennamara, A.; Cherqaoui, A.; Takky, D. (2005) *Applied Surface Science*, Volume 252, Issue 23, Pages 8178–8184.

Abdallah, M. (2002) *Corrosion Science*, 44, 717–728.

Abd-El Rehim, S. S.; Ibrahim, M. A. M.; Khalid, K. F. (2001), *Materials Chemistry and Physics*, 70, 270.

Abed, Y.; Arrar, Z.; Hammouti, B.; Aouniti, A.; Kertit, S.; Mansri, A. (1999) *J. Chim. Phys.*, 96, 1347–1355.

Abiola, O. K.; Oforko, N. C. (2004) *Materials Chemistry and Physics*, 83, 315–322.

ACI Committee Report (1985) "Corrosion of Metals in Concrete," *ACI Journal*.

Adamiec, P.; Dziubinski, J. (2003) "Hydrogen Cracks in Welded Steel Pipes-Part 1: Formation and Parameters," *Welding Research Abroad*, 49, 1, 24–27.

Agrawal, R.; Naamboodhiri, K. T. G. (1990) "The Inhibition of Sulfuric Acid Corrosion of 410 Stainless Steel BY Thioureas," *Corrosion Science*, 30, 1, 37–52.

Akin, S.; Gungor, A.; Akin, M. (2006) "Effect of 1H-1,2,4 Triazole-3-Thiol and borax inhibitors on C-1040 steel materials in tap water," *Proceedings*, 10th International Corrosion Symposium, Adana, Turkey, 197.

Akoz, F.; Yuzer, N.; Koral, S. (1996) "Effect of Magnesium Chloride to Reinforced Concretes Made of Silica Fumes and Without," *Proceedings*, 4th National Concrete Congress, 30–31, 11, 317–326.

Al-Amoudi, O. S. B.; Maslehuddin, M. (1993) *Cement and Concrete Research*, 223, 139–146.

Al-Amoudi, O. S. B.; Maslehuddin, M.; Lashari, A. N.; Almusallam, A. A. (2003) "Effectiveness of Corrosion Inhibitors in Contaminated Concrete," *Cement Concrete Comp.*, 25, 439–449.

Alonso, C.; Andrade, C.; Gonzalez, J. A. (1988) "Relation Between Resistivity and Corrosion Rate of Reinforcements in Carbonated

Mortar Made with Several Cement Types," *Cement and Concrete Research*, 18, 5, 687–698.

Altan, H.; Sen, S. (2004) *Materials and Design*, 25, 637.

Al-Tayyib, A. J.; Khan, M. S. (1988) "Corrosion Rate Measurements of Reinforcing Steel in Concrete by Electrochemical Techniques," *ACI Materials Journal*, May-June issue, 172–177.

Andrade, C.; Alonso, C; Molina, F. J. (1993) "Cover Cracking as a Function of Bar Rebar Corrosion: Part I-Experimental Test," *Materials and Structures*, 26, 453–464.

Anoicai, L.; Masi, R. (2005) *Corrosion Science*, 47, 2883.

Applegate, L. M. (1960) *Cathodic Protection*, McGraw-Hill Books, New York, 90.

Ariel, N. (1981) *Corrosion and Chemical Cleaning in Water-Steam Cycle Units*, Ankara, Turkey, PETKIM publications, 89.

Arslan, M; Aslan, F.; Eflatun, A.; Asan, M. B. (1996) *Effect of Acid Rains on Historical Sites*, Firat University Research and Development Project, FUNAF-74, Elazig, Turkey.

Asan, A.; Kabasakaloglu, M.; Isiklan, M.; Kilic, Z. (2005) *Corrosion Science*, 47, 1534–1544.

Ashassi-Sorkhabi, H.; Shaabani, B.; Sefzadeh, D. (2005) *Electrochimica Acta*, 50, 3446–3452.

Ashworth, V.; Googan, C. (1993) *Cathodic Protection, Theory and Practice*, Ellis Horwood Pub.

*ASM Handbook, Vol. 11, Failure Analysis and Prevention*, ASM International, Materials Park, OH, 1986.

*ASM Handbook, Vol. 19, Fatigue and Fracture*, ASM International, Materials Park, OH, 1996.

ASTM C-876–91, American Society for Testing and Materials, Standard Test Method for Half-Cell Potentials of Uncoated Reinforcing Steel in Concrete.

ASTM G1-90, American Society for Testing and Materials, Standard Practice for Preparing, Cleaning and Evaluating Corrosion Test Specimens.

Ateya, B. G.; El-Anoduili, B. E.; El-Nizamy, F. M. (1993) "The Effect of Thiourea and the Corrosion Kinetics of Mild Steel in H2SO4," *Corrosion Science*, 24, 6, 122–129.

Atkinson, J. T. N; Van Droffelaar, H. (1983) *Corrosion and Its Control*, NACE, Houston.

Awady, A. A.; Abd-El-Nabey, B. A.; Aziz, S. G. (1993) *Journal of the Chemical Society Faraday Transactions*, 84, 795.

Baeckman, W.; Schwenk, W.; Prinz, W. (1997) *Handbook of Cathodic Corrosion Protection-Theory and Practice of Electrochemical Protection Processes*, 3rd Edition.

Banaz, A.; Esener, H.; Asan, A. (2006) *Proceedings*, 10th International Corrosion Symposium, Adana, Turkey, 309.

Batis, E. R. (2005) "Corrosion of Steel Reinforcement Due to Atmospheric Pollution," *Cement and Concrete Research*, 27, 269–275.

Bauman, J. (2004) "Well Casing Cathodic Protection Utilizing Pulse Current Technology," *J. of Corrosion*, 1–4.

Bazzaoui, M.; Martins, J. I.; Costa, S. C.; Bazzaoui, E. A.; Reis, T. C.; Martins, L. (2006) *Electrochimica Acta*, 511, 2417.

Becerra, H. Q.; Retamoso, C.; Macdonald, D. D. (2000) "The Corrosion of Carbon Steel in Oil-in-Water Emulsions Under Controlled Hydrodynamic Conditions," *Corrosion Science*, 42, 561–575.

Beech, I. B.; Sunner, J. (2004) *Current Opinion in Biotechnology*, 15, 181–186.

Benedict, R. L. (1990) "Corrosion Protection of Concrete Cylinder Pipe," *Materials Performance*.

Bensalah, W.; Elleuch, K.; Feki, M.; Wery, M.; Ayedi, H. F. (2007) *Surface Coating Technology*, 201, 7855.

Bentiss, F. Traisnel, M., Lagrenee, M. (2001) *Journal of Applied Electrochemistry*, 31, 41.

Berke, N. S. (1991) *Concrete International*, 24–27.

Bernard, M. C.; Hugot-Le Goff, A.; Joiret, S.; Dinh, N. N.; Toaan, N. N. (1999) *Journal of Electrochemical Society*, 146, 995.

Binici, H.; Kaplan, H.; Yasarer, F. (2009) *Durability of Concrete Coated With Dye Incorporating Pumice*, Institute of Fundamental Sciences, Suleyman Demirel University, Isparta, Turkey, 13, 1, 40–47.

Bocchetta, P.; Sunseri, C.; Bottino, A.; Capannelli, G.; Chivarotti, G.; Piazza, S. (2002) *Journal of Applied Electrochemistry*, 32, 977.

British Standards, 1991, Cathodic Protection, BS 7361, Part-I.

Brondel, D.; Edwards, R.; Hayman, A.; Hill, D.; Mehta, S.; Semerad, T. (1994) "Corrosion in the Oil Industry," *Oilfield Review*.

Brunelli, K.; Dabala, M. (2005) *Corrosion Science*, 47, 989.

Bruno, T. V. (1997) *Causes and Prevention of Pipeline Failures*, Metallurgical Consultant Inc.

Buyuksagis, A. (2007) "Deposits and Corrosion Occurring at Afyonkarahisar Geothermal Heating System," *Journal of Geological Engineering*, 2.

Buyuksagis, A. (2008) "Formation of Deposit and Corrosion in Afyonkarahisar Geothermal Heating System," *Geological Engineering*, 32, 1, 9–24.

Buyuksagis, A.; Erol, S. (2008) "Prevention of Deposits and Corrosion at Afyonkarahisar Geothermal Heating System," *Proceedings*, Thermal and Mine Waters Conference, 249–263.

Cakir, A. F. (1990) "Metallic Corrosion and its Control," *Journal of Mechanical Engineers*, Ankara, Turkey, 131, 128–129, 364.

Cao, C.; Zhang, J. (2002) *An Introduction to Electrochemical Impedance Spectroscopy*, The Science Press, Beijing, China.

Carter, R. M. (1975) *Coatings for Crude Oil Tank Bottoms*, 13, 270.

Ceremoglu, B.; Ozkan, M.; Karadeniz, F. (2003) *Fruit and Vegetable Processing Technology*, Food Technology Institution Publications, 28, Ankara, Turkey.

Champion, F. A. (1952) *Corrosion Testing Procedures*, Chapman and Hall, London.

Cheung, C. W. S.; Beech, I. B. (1996) *Biofouling*, 9, 231–249.

Chang, Y. (1989) *Journal of Material Science*, 24, 14.

Chen, R. Y.; Yuen, W. Y. D. (2003) *Oxidation of Metals*, 59, 433.

Chetouani, A.; Medjahed, K.; Benabadji, K. E.; Hammouti, B.; Kertit, S.; Mansri, A. (2003) *Progress in Organic Coatings*, 46, 312–316.

Cheung, C. W. S.; Beech, I. B.; Campbell, S. A.; Satherley, J.; Schiffrin, D. J. (1994) *International Biodeterioration aand Biodegredation*, 33, 299–310.

Chevron Corporation (1997) *Corrosion Prevention Manual*.

Cil, Ismail (2006) "Corrosion Measurement of Rebars by Electrical Methods," *Science and Engineering Journal*, 9 September, University, Izmir, Turkey

Cleas, W. (1991) Cooling Water Treatment, Technical Report.

Clubley, B. G. (1990) *Chemical Inhibitors for Corrosion Control*, The Royal Society of Chemistry, Newcastle.

Colangelo, V. J.; Heiser, F. A. (1987) *Analysis of Metallurgical Failures*, 2nd edition, Wiley, New York.

Corrosion and Erosion Management (2003) Under Balanced Drilling (UBD) Operations, International Association of Drilling Contractors version 1.5, retrieved at http://www.iadc.org/committees/underbalanced/Documents/IADC%20%20Corrosion%20Management1.5.pdf.

Craig, B. D.; Anderson, D. (1995) *Handbook of Corrosion Data*, 2nd ed. ASM International, Materials Park, OH.

Crolet, J. L. (2000) *Corrosion Science*, 42, 1303–1304.

Cruz, J.; Martinez, R.; Genesca, J.; Ochoa, E.G. (2004) *J. Electroanalytical Chemistry*, 566, 111.

Dag, E.; Koc, T.; Pamuk, V. (1996) "Corrosion Protection of Polyurethane Coated Steel Pipelines Used in Geothermal Central Heating System of Afyon City," *Proceedings*, 5$^{th}$ Corrosion Symposium, Adana, Turkey, 349–357.

Damborenea, J. De; Bastidas, J. M.; Vazquez, A. J. (1997) *Electrochimica Acta*,42, 455.

Davis, J. R. (ed.) (1987) *Metals Handbook*, 9th edition, ASM International, Metals Park, OH, 13.Dehri, I; Erbil, M. (1996) "The Corrosion of Aluminum in NH3 and KOH Solutions at the Same pH and the Effect of Sulfate Ions," *Proceedings*, 5$^{th}$ Corrosion Symposium, Adana, Turkey

Doruk, M. (1972) *Electrochemical Principles of Corrosion*, Middle East Technical University Publications, Ankara, Turkey.

Doruk, M. (1982) *Corrosion and Its Prevention*, Middle East Technical University Publications, Ankara, Turkey, 70, 41.

Doruk, M. (1982) *Corrosion Protection*, Middle East Technical University Publications, Ankara, Turkey.

Doruk, M. (1996) "Principles of Corrosion," *Proceedings*, 5$^{th}$ Corrosion Symposium, Adana, Turkey.

Ebunso, E. E.; Ekpe, U. J.; Ita, B. I.; Offiong, O. E.; Ibok, U. J. (1999) *Materials Chemistry and Physics*, 60, 79–90.

Eflatun, A. (1994) *Effect of Acid Rains on Historical Artifacts*, Master Thesis, Firat University, Elazig, Turkey.

Eksi, A. (1976) *Reasons, Results and Prevention Methods of Corrosion in Cans*, Food Quality Control Education and Research Institution Publications, 6, 43, Bursa, Turkey. Eminoglu, F. (1996) "Treatment of Boiler System Waters," *Proceedings*, Henkel Co., 5$^{th}$ Corrosion Symposium, Adana, Turkey, 282–288.

Emregul, K. C.; Hayvali, M. (2006) *Corrosion Science*, 48, 797–812.

Erbil, M. (1971) *Corrosion Rate Determination of Aluminum in Lactic Acid Solution*, Master Thesis, Ankara University Physical Chemistry Division.

Erbil, M. (1975) *Aluminum Corrosion in Various Aqueous Solutions*, PhD Dissertation, Ankara University, Arts and Sciences Faculty, Ankara.

Erbil, M. (1975) "Aluminum's Passivity in Na2SO4 Solutions and Its Corrosion in Aqueous Solutions Studied by Current-Potential Curves," *Proceedings*, 5th Science Congress, TUBITAK, Ankara, 397–413.

Erbil, M. (1980) *Investigation of Novel Inhibitors of Iron Corrosion*, Ankara University, Faculty of Arts and Sciences, Ankara, Turkey.

Erbil, M. (1984) *Corrosion Inhibitors and Determination of Inhibitor Efficiencies*, SEGEM Press, Ankara.

Erbil, M. (1996) "Corrosion Protection," *Proceedings*, 5th Corrosion Symposium, Adana, Turkey, 223–237.

Erbil, M.; Uneri, S. (1974) "The Linear Polarization Measurements for Aluminum in Organic Acidic Media," *Chimica Acta Turcica*, 2, 30–41.

Erbil, M.; Yazici, B.; Yilmaz, A. B. (1996) "The Corrosive Effects of Magnesium Ion on the Reinforcement Steel and Concrete," *Proceedings*, 5th Corrosion Symposium, Adana, Turkey, 360–368.

Erdogan, T. Y. (2003) *Concrete*, Middle East Technical University Development Publications, Ankara, Turkey.

Ergun, M. (1982) *A Study of Aluminum's Passivity via Potentiostatic Technique*, PhD Dissertation, Ankara University, Arts and Sciences Faculty, Ankara.

Erol, S. (2008) *Deposits and Corrosion Problems at Afyonkarahisar Geothermal Heating System and Solutions to Problems*, Master's Thesis, Afyon Kocatepe University Institute of Fundamental Sciences, Afyon, Turkey.

Ersoy, U. (1987) *Basic Principles of Reinforced Concrete and Carrying Loads*, Evrim Publications, Istanbul, 1, 643.

Eryurek, I. B. (1993) *Damage Determination*, Faculty of Mechanical Engineering, Istanbul Technical University, Istanbul, Turkey, 112.

Euro Inox (2005) *Stainless Steel Surfaces Manual*, Building Series, 1, Luxembourg.

European Federation of Corrosion Publications (1995) *Guidelines on Materials Requirements for Carbon and Low Alloy Steels for H2S Containing Environments in Oil and Gas Production*, the Institute of Materials, 16.

Fattuhi, N. I.; Hughes, B. P. (1998) *Journal of Concrete Research*, 40, 159–166.

Feldman, R. F.; Cheng-yi, H. (1985) "Resistance of Mortars Containing Silica Fume to Attack by a Solution Containing Chlorides," *Cement and Concrete Research*, 15, 3, 411–420.

Flick, E. W. (1987) *Corrosion Inhibitors*, Noyes Publications, Park Ridge, New Jersey.

Fonseka, I. T. E.; Cristina, A.; Marin, S. (1992) *Electrochemica Acta*, 37, 13, 2541–2548.

Fontana, M. G. (1987) *Corrosion Engineering*, McGraw-Hill Book Company, Singapore.

Fontana, M. G. (1986) *Corrosion Engineering*, 3rd ed., McGraw-Hill, New York.

Fontana, M. G.; Greene, M. D. (1967) *Corrosion Engineering*, McGraw-Hill Book Co., New York.

Fontana, M. G.; Greene, N. D. (1978) *Corrosion Engineering*, 2nd edition, McGraw-Hill, NY.

Fontana, M. G.; Greens, N. D. (1967) *Corrosion Engineering*, McGraw-Hill Inc., New York, NY.

Franco, R. J. (1995) "Materials Selection for Produced Water Injection Piping," *Materials Performance*.

Fultz, B. S. (1988) "Coatings With and Without Cathodic Protection in Salt Water Ballast Tanks," *Material Performance*, 24–30.

Gad-Allah, A. G.; Abou-Romia, M. M.; Badawy, M. W.; Rehan, H. H. (1991) *J. Appl. Electrochem.*, 21, 829.

Galip, H.; Erbil, M. (1992) "The Effect of SO42- Ions on the Corrosion of Iron in Artificial Seawater," *Journal of Chemistry*, 16, 268–272.

Gaylarde, C. C.; Johnston, J. M. (1983) *Proceedings*, Prof. Conf. Microbial Corrosion, The National Physical Laboratory and The Metal Society, NPL Teddington, 127.

Gendenjamts, O.E. (2005) Interpretation of Chemical Composition of Geothermal Fluids from Arskogsstrond, Dalvik and Hrisey, Iceland and Khangai, Mongolia, Geothermal Training Program, The United Nations University, Report # 10, Reykjavik, Iceland.

Gibala, R.; Hehemann, R. F. (1984) *Hydrogen Embrittlement and Stress Corrosion Cracking*, ASM International, Materials Park, OH.

Ginzburg, V. B. (1989) *Steel-Rolling Technology: Theory and Practice*, Marcel Dekker, New York, NY.

Girginov, A.; Popova, A.; Kanazirski, I.; Zahariev, A. (2006) *Thin Solid Films*, 515, 1548.

Girit, N. (2003) "Corrosion Inhibitors," *Engineering News*, Turkey, 4, 426, 139

Gonzalez, J. A.; Andrade, C.; Alonso, C.; Feliu, S. (1995) "Comparison of Rates of General Corrosion and Maximum Pitting Penetration on Concrete Embedded Steel Reinforcement," *Cement and Concrete Research*, 25, 2, 257–264.

Grabke, H. J.; Leroy, V.; Viefhaus, H. (1995) *ISIJ International*, 35, 95.

Graf, M.; Grimpe, F.; Liessem, A.; Popperling, R. K. (1999) Review of the HIC Test Requirements for Linepipe Over the Years of 1975 to 2000, Europipe, Germany.

Green, M. M. (1983) "Cathodic Protection Design," *A. F. Manual* 88–45, Washington.

Guimard, N. K.; Gomez, N.; Schmidt, C. E. (2007) *Prog. Polymer Science*, 32, 876.

Gulikers, J. J. W.; Mier, J. G. M. (1991) "The Effect of Pateh Repairs on the Corrosion Rate of Steel Reinforcement in Concrete," *Proceedings*, II. ACI International Conference on Durability of Concrete, Montreal, Canada, Supplementary Papers, 445–460.

Gurten, A. A. (2002) *Effects of Polyvinylpyrolidone and thiosemicarbazide on Corrosion of Reinforced Steel*, Nigde University Institute of Fundamental Sciences, Nigde, Turkey.

Hamid, A. A.; Yahya, R. K. (2003) "Influence of Fretting on the Fatigue Strength at the Viseclamp-Specimen Interface, Indian Academy of Sciences," *Bulletin of Material Science*, 226, 7, 749–754.

Hansson, C. M. (1984) "Comments on Electrochemical Measurements of the Rate of Corrosion of Steel in Concrete," *Cement and Concrete Research*, 14,4, 574–584.

Hansson, C. M.; Frolund, T.; Markussen, J. B. (1985) "The Effect of Chloride Cation Type on the Corrosion of Steel in Concrete by Chloride Salts," *Cement and Concrete Research*, 15, 1, 65–73.

Haque, M. N.; Kayyali, O. A. (1995) "Free and Water Soluble Chlorides in Concrete," *Cement and Concrete Research*, 25, 3, 531–542.

Hasanov, R.; Sadikoglu, M; Bilgic, S. (2007) *S. Appl. Surf. Sci.*, 253, 3913.

Hausman, D. A. (1964) "Electrochemical Behavior of Steel in Concrete," *ACI Journal*, 61, 2, 170–187

Hertzberg, R. W. (1996) *Deformation and Fracture Mechanics of Engineering Materials*, 4th edition, Wiley, New York.

Heywood, R. B. (1962) *Designing against Fatigue*, Chapman and Hall Ltd., London, U.K.

Hiao, H.; Chih, H.; Tai, W. (2005) *Surface and Coatings Technology*, 199, 127.

Hirsch, G. P. (1974) *Werkstoffe und Korrosion*, 25, 10.

Hoar, T. P. (1961) "Electrochemical Principles of the Corrosion and Protection of Metals," *J. Appl. Chem.*, 11, 121.

Honeycombe, R.; Bhadeshia, H. K. D. H. (1995) *Steels' Microstructure and Properties*, 2nd edition, Edward Arnold, London, U.K.

Hossenia, M.; Mertens, S. F. L.; Ghorbanic, M.; Arshadi, M. R. (2003) *Materials Chemistry and Physics*, 78, 800–808.

Hurst, C. J. (2002) *Manual of Environmental Microbiology*, ASM Press, U.S., 1138.

Iroh, J. O.; Su, W. (1997) *Journal of Applied Polymer Science*, 71, 2075.

Iroh, J. O.; Su, W. C. (2000) *Electrochim. Acta*, 46, 15.
Jagminas, A.; Lichusina, S.; Kurtinaitiene, M.; Selskis, A. (2003) *Applied Surface Science*, 211, 194.
Johnson, S. D. (1988) "FRP Linings for Tank Bottom Repair," *Material Performance*, 32–34.
Jones, D. A. (1992) *Principles and Prevention of Corrosion*, MacMillan Pub. Co, New York.
Jones, D. A. (1996) *Principles and Prevention of Corrosion*, 2nd ed. Pearson Education, Upper Saddle River, NJ.
Jones, L. W. (1992) *Corrosion and Water Technology for Water Producers*, OGCI Publications, Oil and Gas Consultants International Inc., 2nd Edition, Tulsa, OK.
Kahyaoglu, H. (1998) *Corrosion Chemistry*, PhD Dissertation, Cukurova University, Institute of Fundamental Sciences, Adana, Turkey.
Karavaizoglu, N. (1996) "Cathodic Protection Application on Water Pipelines Made of Steel," *Proceedings*, 5th Corrosion Symposium, Adana, Turkey, 340–343.
Katoh, M. (1968) *Korrosion Science*, 8, 423.
Kayali, O.; Zhu, B. (2005)" Corrosion Performance of Medium-Strength and Silica Fume High-Strength Reinforced Concrete in a Chloride Solution," *Cement and Concrete Composites*, 27, 117–124.
Kayyali, O. A.; Haque, M. N. (1995) "The Cl–/OH– Ratio in Chloride Contaminated Concrete-A Most Important Criterion," *Journal of Concrete Research*, 47, 172, 235–242.
Keijman, J. M. (1997) "The Use of Novel Siloxane Hybrid Polymers in Protective Coatings," *Oil Gas Eur Mag*, 23(3), 38–40.
Kelestemur, O. (2008) *Investigation of Applicability and Corrosion Resistivity of Dual-Phase Steel in Reinforced Concretes*, PhD dissertation, Firat University Institute of Fundamental Sciences, Elazig, Turkey.
Kelestemur, O.; Yildiz, S. (2005) "The Effect of Styrofoam on the Corrosion of Steel Reinforced Concrete," *Journal of Olytechnic*, 8, 3, 311–315.
Keresztes, Z.; Felhosi, I.; Kalman, E. (2001) *Electrochimica Acta*, 46, 3841–3849.
Kertit, S.; Hammouti, B. (1996) *Applied Surface Science*, 93, 59.
Khan, M. S. (1991) "Corrosion State of Reinforcing Steel in Concrete at Early Ages," *ACI Materials Journal*, 88, 1, 37–40.
Kiker, B. (2004) *Corrosion and Mechanical Wear on Equipment Used in Handling Produced Water*, Petroleum Technology Transfer Council (PTTC), OK.

Kikuchi, T.; Aramaki, K. (2000) "The Inhibition Effects of Anion and Cation Inhibitors on Corrosion of Iron in an Anhydrous Acetonitrile Solution of FeCl3," *Corrosion Science*, 42, 817–829.

Kleikemper, J.; Schroth, M. H.; Sigleer, W. V.; Schmucki, M.; Bernasconi, S. M.; Zeyer, J. (2002) *Applied and Environmental Microbiology*, 68, 1516–1523.

Klimartin, P. A.; Trier, L.; Wright, G. A. (2002) *Synthetic Metals*, 131, 99–109.

Ko, S.; Lee, D.; Jee, S.; Park, H.; Lee, K.; Hwang, W. (2006) Thin Solid Films, 515, 1932.

Kouloumbi, N; Batis, G. (1992) "Chloride Corrosion of Steel Rebars in Mortars with Fly Ash Admixtures," *Cement and Concrete Composites*, 14, 199–207.

Kucera, J.; Hadjuga, M.; Glowacki, J.; Broz, P. Z. (1999) *Z. Metall.*, 90, 514.

Kumar, S.; Kameswara, C. V. S. R. (1994) *Cement and Concrete Research*, 24, 1237–1244.

Kurtkaya, N. (2008) "Pipelines and Energy Bridges," *Proceedings*, XI. International Corrosion Symposium, International Corrosion Symposium, Izmir, Turkey.

Kustu, C.; Emregul, K. C.; Hayvali, M. (2006) "Effect of New Schiff Base Synthesized From Phenazone and Vanillin on Steel Corrosion that is in 2 M HCl Solution," *Proceedings*, 10th International Corrosion Symposium, Adana, Turkey, 330.

La Que, F. L. (1975) *Marine Corrosion Causes and Prevention*, John Wiley and Sons, New York.

Lagranee, M.; Mernari, B.; Bouanis, M.; Traisnel, M.; Bentiss, F. (2002) *Corrosion Science*, 44, 573–588.

Larabi, L.; Harek, Y. (2004) *Portugaliae Electrochimica Acta*, 22, 227–247.

Lee, H. P.; Nobe, K. (1986) *J. Electrochem. Soc.*, 133, 2035.

Lee, W.; Characklis, W. G. (1993) *Corrosion*, 49, 186–199.

Lee, W.; Lewandowski, Z.; Morrison, M.; Characklis, W. G.; Avci, R.; Nielsen, P. H. (1993) *Biofouling*, 7, 217–239.

Lewis, D. A.; Copenhagen, W. J. (1959) "Corrosion of Reinforcing Steel in Concrete in Marine Atmospheres," *Corrosion*, NACE, 15.

Li, J. (2003) "Fe-Ti Composite Powder and Its Application in Anti-Corrosive Coatings," *Coating Industry*, 33, 7, 51–53.

Liam, K. C.; Ray, S. K.; Nortwood, D. O. (1992) "Chloride Ingress Measurement and Corrosion of Potential Mapping Study of a 24 year old Reinforced Concrete Jetty Structure in a Tropical Marine Environment," *Journal of Concrete Research*, 44, 160, 205–215.

Liu, G. (1995) "Flourocarbon Coatings with Super Weatherability," *China Coatings*, 3, 21–26.

Locke, C. E.; Riggs, O. L. (1970) *Anodic Protection, Theory and Practice in Prevention of Corrosion*, Plenium Press, London.

Loreer, P.; Lorenz, W. J. (1980) "The Kinetics of Iron Dissolution and Passivation in Solutions Containing Oxygen," *Electrochemica Acta*, 25, 375–381.

Lorenz, W. J.; Mansfeld, F.; Kendig, M. W. (1984) *Proceedings*, International Congress on Metallic Corrosion, 14.

Lou, H.; Guan, Y. C.; Han, K. N. (1998) *Corrosion*, 54, 9.

Luo, H.; Guan, Y. C.; Han, K. N. (1998) *Corrosion*, 54/8, 619–627. Madayag, A. F. (1969) *Metal Fatigue: Theory and Design*, John Wiley and Sons Inc., New York

Magot, M.; Ollivier, B.; Patel, B. K. C. (2000) *Antonie van Leeuwenhook*, 77, 103–116.

Marcus, P.; Oudar, J. (1995) *Corrosion Mechanisms in Theory and Practice*, Marcel Dekker, New York.

Mark, P.; Piatak, M. A.; Huadmai, J.; Dias, G. (2006) *Biomaterials*, 27, 1728.

Marks, M. F. (2004) *Materials Letters*, 58, 3316.

Masadeh, S.; Doruk, M. (1996) "Embrittlement in Pre-stressed Bars Due to Cathodically Evolved Hydrogen Gas in Solutions Similar to Concrete Environment," *Proceedings*, 5th Corrosion Symposium, Adana, Turkey, 194.

Mattson, E. (1989) *Basic Corrosion Technology for Scientists and Engineers*, John Wiley and Sons, New York

McNaughton, K. J. (1980) *Materials Engineering Volume I: Selecting Materials for Process Equipment*, McGraw-Hill Book Co., New York.

McNaughton, K. J. (1980) *Materials Engineering Volume II: Controlling Corrosion in Process Equipment*, McGraw-Hill Book Co., New York.

Mechanism of Corrosion Steel in Concrete (1985) "ACI Committee Report 222," *ACI Journal*.

Mehta, P. K. (1986) *Concrete, Structure Properties and Materials*, Prentice-Hall, U.S., 152–159.

Mergen, H.; Mergen, B. E.; Erdogmus, A. B. (2006) "Geothermal Energy Applications," *J. PMR Science*, 9, 3, 108–113.

Migahed, M. A. (2005) *Materials Chemistry and Physics*, 93, 48.

Mitra, A. K. (1996) *Research and Development Journal*, NTPC 2, 52.

Mitra, S. (2002) "Basic Corrosion Control Well Casing Protection," presentation notes, Corpro.

Morgan, J. (1987) *Cathodic Protection*, NACE, 2nd edition, Houston, TX.

Morgan, J. H. (1985) *Cathodic Protection*, NACE, Houston.

Moutarlier, V.; Gigandet, M. P.; Pagetti, J.; Ricq, L. (2003) *Surface Coating Technology*, 173, 87.

Mozalev, A.; Poznyak, A.; Mozaleva, I.; Hassel, A. W. (2001) *Electrochemistry Communications*, 3, 299.

Munger, C. G. (1984) *Corrosion Prevention by Protective Coating*, NACE, Houston.

Mutlu, H. (1997) "Geochemical Characteristics of Afyon Gazligol Thermal and Mine Waters," *Journal of Geological Engineering*, 50.

NACE (National Association of Corrosion Engineers) (1979) Corrosion Control and Water Technology, Technical Report, publication no. TPC.5, Houston, TX.

National Association of Corrosion Engineers (NACE) (1975) Control of Internal Corrosion in Steel Pipelines and Piping Systems, Report #0175, p. 75.

National Association of Corrosion Engineers (NACE) (1993) "Developments in Corrosion Monitoring Processes," *Proceedings*, International Corrosion Congress, Houston, TX, 3A.

National Association of Corrosion Engineers (NACE) International (2007) CP1 Cathodic Protection Tester, Course Manual.

Neal, M. J.; Gee, M. (2001) *Wear Problems and Testing for Industry*, William Andvew Publishing, U.K.

Neville, A. M. (1987) *Properties of Concrete, Longman Science and Technology*, Longman Group, Longman House, Essex, U.K.

Oswin, H. G.; Cohen, M. (1975) "Study of the Cathodic Reduction of Films on Iron," *Journal of the Electrochemical Society*, 104, 1, 9–16.

Ozbas, M. (1997) *Corrosion Protection via Suitable Metal Selection and Design*, Gazi University Institute of Fundamental Sciences, Ankara, Turkey, 126.

Ozdemir, I. H. (1981) *General Inorganic and Technical Chemistry*, Press Technicians Press, Istanbul, Turkey.

Oztan, Y. (1995) *Air Pollution, Environmental Pollution*, Trabzon, Turkey.

Parker, M. E. (1967) *Pipeline Corrosion and Cathodic Protection*, NACE, Houston.

Patermarakis, G.; Moussoutzanis, K. (2003) *Chemical Engineering Communications*, 190, 1018.

Patton, C. C. (1993) "Corrosion Control in Water Injection Systems," *Materials Performance*, 32, 46–49.

Patzay, G.; Stahl, G.; Karman, F. H.; Kalman, E. (1998) "Modeling of Scale Formation and Corrosion from Geothermal Water," *Electrochim Acta*, 43, 137–147.

Peabody, A. W. (1967) *Control and Pipeline Corrosion*, NACE, Houston.

PETKIM (Turkish Governmental Petrochemical Industry) (1988) *Refining Equipments Control and Repair Manual*, PETKIM Publications, 03.

Popova, A.; Sokolova, E.; Raicheva, S.; Christov, M. (2003) *Corrosion Science*, 45, 33.

Popovic, M. M.; Grgur, B. N.; Miskovic-Stankovic, V. B. (2005) *Prog. Org. Coat.*, 52, 359.

Pourbaix, M. (1966) *Atlas of Electrochemical Equilibria in Aqueous Solutions*, Pergamon Press, London, U.K.

Pourbaix, M.; Zoubov, N. (1966) *Atlas of Electrochemical Equations in Aqueous Solutes*, J. W. Arrowsmith Ltd., U.K.

Quaraishi, M. A.; Ansari, F. A.; Jamal, D. (2002) *Material Chemistry and Physics*, 687–690.

Quartorone, G.; Bonaldo, L.; Tartato, C. (2006) *App. Surf. Sci.*, 252, 8251.

Quraishi, M. A.; Jamal, D. (2003) *Materials Chemistry and Physics*, 78, 608–613.

Quraishi, M. A.; Rawat, J.; Ajmal, M. (1998) *Corrosion Science Section*, 54, 12, 996–1002.

Rahmel, A.; Tobolski, J. (1965) *Corrosion Science*, 5, 333.

Rasheeduzzafar, D. F. H.; Bader, M. A.; Khan, M. M. (1992) "Performance of Corrosion Resisting Steel in Chloride-Bearing Concrete," *ACI Materials Journal*, 89, 5, 439–448.

Ren, J.; Zuo, Y. (2004) *Surface Coating Technology*, 182, 237.

Revie, R. W. (2000) *Uhlig's Corrosion Handbook*, 2nd ed., John Wiley&Sons, New York.

Reynolds, R. (2003) Produced Water and Associated Issues, Geological Survey Open-File Report, Report of Petroleum Technology Transfer Council (PTTC), OK.

Richter, S.; Thorarinsdottir, R. I.; Jonsdottir, F. (2007) *Corrosion Science*, 49, 1907.

Roberge, P. R. (1999) *Handbook of Corrosion Engineering*, McGraw-Hill, New York, NY.

Robert, G. K.; John, R. S.; David, W. S.; Rudoplh, G. B. (2003) *Electrochemical Techniques in Corrosion Science and Engineering*, Marcel Dekker, New York, NY, 302.

Rodriguez, P.; Ramirez, E.; Gonzalez, J. A. (1994) "Methods for Studying Corrosion in Reinforced Concrete," *Journal of Concrete Research*, 46, 167, 81–90.

Rozenfeld, L. L. (1981) *Corrosion Inhibitors*, McGraw-Hill Inc., New York, NY.

Rudd, A. L.; Breslin, C. B.; Maansfeld, F. (2000) *Corrosion Science*, 42, 275.

Sahlaoui, H.; Makhlouf, K.; Sidhom, H.; Philibert, J. (2004) *Material Science and Engineering*, A372, 98.

Saidman, S. B. (2002) *Journal of Electrochemistry*, 534, 39.

Sakarya, M. (1996) "Chemical Cleaning of Boilers," *Proceedings*, 5th Corrosion Symposium, Adana, Turkey, 272–276.

Sarayli, M. A. (1978) *Science of Structural Materials*, Kutulmus Press, Istanbul, Turkey

Sarioglu, F.; Javaherdashti, R.; Aksoz, N. (1997) *Int. J. Pres. Ves. and Piping*, 73, 127–131.

Schweitzer, P. A. (1999) *Atmospheric Degradation and Corrosion Control*, Marcel Dekker, New York.

Schweitzer, P. A. (1989) *Corrosion and Corrosion Protection Handbook*, 2nd ed., Marcel Dekker, New York.

Scully, J. C. (1966) *The Fundamentals of Corrosion*, Pergamon Press, Oxford.

Searles, J. L.; Gouma, P. I.; Buchheit, R. G. (2003) "Stress Corrosion Cracking of AA5083 Alloy," *Metall. Mater. Trans. A*, 32, pp. 2859–2867.

Sharp, J. W.; Figg, J. W.; Leeming, M. B. (1988) "The Assessment of Corrosion of the Reinforcement in Marine Concrete by Electrochemical and Other Methods," *Proceedings*, II. International Conference on Performance of Concrete in Marine Environments, New Brunswick, Canada, 21–26, 8, 105–125.

Shokry, H.; Yuasa, M.; Sekine, I.; Issa, R. M.; El-Baradie, H. Y.; Gomma, G. K. (1998) *Corrosion Science*, 40, 2173–2186.

Shreir, L. L. (1976) *Corrosion*, Newnes-Butter Worths, London, U.K., 1&2.

Shreir, L. L. (1977) *Corrosion Volume I and II*, Newnes Butterwords.

Shreir, L. L.; Jarman, R. A.; Burnstein, G. T. (eds.) (1994) *Corrosion*, 3rd Edition, Butterworths-Heinemann, 1–2.

Shreir, L. L.; Jarman, R. A.; Burnstein, G. T. (eds.) (1994) *Corrosion, Metal and Environment Reactions*, Butterworths-Heinemann, 8, 10.

Sierra, R. C.; Sosa, E.; Oropeza, M. T.; Gonzales, I. (2000) "Electrochemical Study on Carbon Steel Corrosion Process in Alkaline Sour Media," *Electrochimica Acta*, 47, 2149–2158.

Silva, J. E. P.; Torresi, S. I. C.; Torresi, R. M. (2005) *Corrosion Science*, 47, 811–822.

Sing, I. (1993) "Inhibition of Steel Corrosion by Thiourea Derivatives," *Corrosion*, 49, 6, 473–478.

Skolnik, L. M. (1974) *Methodology of Fatigue Experiments*, Mashinosreyenie, Moscow, Soviet Union.

Smitha, G.; Iroh, O. (2001) *Polymer*, 42, 9665.

Song, G.; Atrens, A. (1997) *Corrosion Science*, 39, 855.
Song, G.; John, D. H. (2005) *Materials and Corrosion*, 56, 15.
Song, Y. K.; Mansfeld, F. (2005) *Materials and Corrosion*, 56, 229.
Sosa, E.; Sierra, R. C.; Garcia, I.; Oropeza, M. T.; Gonzales, I. (2002) "The Role of Different Surface Damages in Corrosion Process in Alkaline Sour Media," *Corrosion Science*, 44, 1515–1528.
Sourmail, T. (2001) *Material Science and Technology*, 17, 1.
Speller, F. N. (1951) *Corrosion Causes and Prevention*, McGraw-Hill Book Co., New York.
Stern, M.; Geary, A. L. (1957) "Electrochemical Polarization," *J. of Electrochemical Society*, 104, 56.
Stern, M.; Weisert, E. D. (1959) "Experimental Observations on the Relation between Polarization Resistance and Corrosion Rate," *Proc. American Society of Testing Materials*, 59, 1280.
Stevens, R. I.; Fatemi, A.; Stevens, R. R.; Fuchs, F. O. (2000) *Metal Fatigue in Engineering*, 2nd edition, Wiley, New York.
Stupnisek-Lisac, E.; Gazivoda, A.; Madzarac, M. (2002) *Electrochimica Acta*, 47, 4189.
Tabrizi, M. R.; Lyon, S. B., Thompson, G. E.; Ferguson, J. M. (1991) "The Long-Term Corrosion of Aluminum in Alkaline Media," *Corrosion Science*, 32, 7, 733–742
Talbot, D.; Talbot, J. (1998) *Corrosion Science and Technology*, CRC Press, Boca Raton, FL.
Tallman, D. E.; Spinks, G.; Dominis, A.; Wallace, G. G. (2002) *Journal of Solid State Electrochemistry*, 6, 73.
Tan, C. K.; Blackwood, D. J. (2003) *Corrosion Science*, 45, 3, 545–557
Thompson, G. E.; Habazaki, H.; Shimizu, K.; Sakairi, M.; Skeldon, P.; Zhou, X.; Wood, G. C. (1999) *Aircraft Engineering and Aerospace Technology*, 71, 3, 228–238.
Tiller, A. K. (1983) *Proc. Conf. Microbial Corrosion*, The National Physical Laboratory and The Metal Society, NPL Teddington, 127.
Tomashov, N. D. (1966) *Theory of Corrosion and Protection of Metals*, MacMillan, NY.
Toprak, S.; Erbil, M.; Yazici, B. (2006) "Physicochemical Corrosion Behavior of Oils Protecting Steel Materials that Vary by Time," *Proceedings*, 10th International Corrosion Symposium, 502–512.
Trabelsi, W.; Dhouibi, L.; Matoussi, F.; Triki, E. (2004) *Synthetic Metals*, 151, 1, 19–24.
Truong, V. T.; Lai, P. K. (2000) *Synthetic Metals*, 110, 7.

Tsangaraki-Kaplanoglou, I.; Theohari, S.; Dimogerontakis, T.; Kallithrakas-Kontos, N.; Wang, Y. M.; Kuo, H. H.; Kia, S. (2006) *Surface Coating Technology*, 200, 3969.

Tsangaraki-Kaplanoglou, I.; Theohari, S.; Dimogerontakis, T.; Kallithrakas-Kontos, N.; Wang, Y. M.; Kuo, H. H.; Kia, S. (2006) *Surface Coating Technology*, 201, 2749.

Tuken, T.; Arslan, G.; Yazici, B.; Erbil, M. (2004) "The Corrosion Protection of Mild Steel by polypyrrole/polyphenol Multilayer Coating," *Corrosion Science*, 46, 11, 2743–2754

Tuken, T.; Ozyilmaz, A. T.; Yazici, B.; Kardas, G.; Erbil, M. (2004) "Polypyrrole and Polyaniline Top Coats on Nickel Coated Mild Steel," *Progress in Organic Coatings*, 51, 27–35.

Tüken, T.; Yazıci, B.; Erbil, M. (2007) "Polypyrrole modified nickel coating on mild steel," *Materials and Design*, vol. 28 issue 1, p. 208–216.

Tüken, T.; Yazıcı, B.; Erbil, M. (2006) "The use of polyindole for prevention of copper corrosion," *Surface & Coatings Technology*, vol. 200, issue 16–17, p. 4802–4809.

Ucuncu, M. (2000) *Food Packaging*, Aegean University Publications, Izmir, Turkey.

Uhlig, H. H. (1971) *Corrosion and Corrosion Control*, 2nd edition, John Wiley and Sons Inc., New York, NY.

Uhlig, H. H., (1971) *The Corrosion Handbook*, John Wiley and Sons, New York, NY

Uhlig, H. H.; Revie, R. W. (1985) *Corrosion and Corrosion Control*, 3rd ed., John Wiley and Sons, New York.

Uhling, H. H. (1967) *Corrosion and Corrosion Control*, John Wiley and Sons Inc., New York, NY.

Uluata, A. R. (1979) *Concrete as Building Material*, Faculty of Agriculture, Ataturk University, Erzurum, Turkey.

Uneri, S. (1998) *Corrosion and Its Protection*, Corrosion Institute Publications, Ankara, Turkey, 4113.

Uneri, S. (1998) *Korozyon ve Onlenmesi*, Segem, Ankara.

Uomoto, T.; Misra, S. (1988) "Behavior of Concrete Beams and Columns in Marine Environment when Corrosion of Reinforcing Bars Takes Place," *Proceedings*, II. International Conference on Performance of Concrete in Marine Environment, New Brunswick, Canada, 21–26, 8, 127–146.

Videla, H. A. (2002) *International Biodeterioration and Biodegradation*, 49, 259–270.

Von Wolzogen Kuhr, C. A. H.; Vlugt van der, L. S. (1934) *Water*, 18, 147–165.

Vracar, L.; Guidi, G. G. (2002) *Corrosion Science*, 44, 1669–1680.

Wahdan, M. H.; Hermas, A. A.; Morad, M. S. (2002) *Materials Chemistry and Physics*, 76, 111–118.

Wang, H. L.; Liu, R. B.; Xin, J. (2004) *Journal of Corrosion Science*, 46, 2455–2466.

Waterhouse R. B. (1972) *Fretting Corrosion*, Pergamon Press, Oxford, U.K.

Wernick, S.; Pinner, R.; Sheasby, P. G. (2001) *The Surface Treatment and Finishing of Aluminum and Its Alloys*, ASM International, Metals Park, OH.

West, J. M. (1970) *Electrodeposition and Corrosion Processes*, 2nd edition, Van Nostrand, London, U.K.

Witte, F.; Knese, V. (2005) *Biomaterials*, 26, 3557.

Wranglen, G. (1987) *An Introduction to Corrosion and Protection of Metals*, Chapman and Hall, New York.

Wulpi, D. J. (1999) *Understanding How Components Fail*, 2nd edition, ASM International, Materials Park, OH.

Xiao-Ci, Y.; Hong, Z.; Ming-Dao, L.; Hong-Xuan, R.; Lu-An, Y. (2000) *Corrosion Science*, 42, 645.

Yagan, A.; Pekmez, N. O.; Yildiz, A. (2006) *Electrochim Acta*, 51, 2949.

Yalcin, H.; Koc, T. (1991) *Corrosion of Iron and Steel Structures and Their Cathodic Protection*, General Directorate of Bank of Cities Publications, Ankara, Turkey, 327.

Yalcin, H.; Koc, T. (1999) *Cathodic Protection*, Palme Publications, Turkey, 356.

Yalcin, H.; Koc, T. (2004) Corrosion and Corrosion Prevention of Reinforced Concretes, CMS Congress Management Systems International Publication and Communication Services Ltd. Co., Ankara, Turkey.

Yalcin, H.; Koc, T. (2004) Corrosion and Corrosion Prevention of Reinforced Steels Embedded in Concrete, CMS Congress Management Systems International Organization, Publication and Communication Services Ltd., Ankara, Turkey, 270.

Yalcin, H.; Koc, T. (2004) Corrosion of Reinforced Steel Bars and Their Prevention, CMS Congress Management Systems International Organization, Publication and Communication Services Ltd., Ankara, Turkey.

Yalcin, H.; Koc, T.; Hoshan, P. (1994) "Corrosion Measurement of Reinforced Concrete Steels via Three Electrode Method," *Proceedings*, 4th Corrosion Symposium, 133–144.

Yalcin, H.; Koc, T.; Pamuk, V. (1996) "Corrosion Prevention of Inner Surfaces of Crude Oil Storage Tanks via Cathodic Protection

Using Dual Galvanic Anode System," *Proceedings*, 5th Corrosion Symposium, Adana, Turkey, 297.

Yano, J.; Nakatani, K.; Harima, Y.; Kitani, A. (2007) *A. Mater. Lett.*, 61, 1500.

Yilmaz, A. B.; Dehri, I.; Erbil, M. (2003) "Effect of pH Change in Admixtures on Corrosion Potential of Reinforced Concrete and on Its Strength," *Proceedings*, 4th Electrochemistry Days, 146–149.

Yurttutan, I. (1996) "Reasons, Monitoring and Prevention of Internal Corrosion Occurring at Adiyaman Petroleum Pipelines," *Proceedings*, 5th Corrosion Symposium, Adana, Turkey, 175–176.

Zang, T.; Zeng, C. L. (2005) *Electrochimica Acta*, 50, 24, 4721–4727.

Zemajtis, J. (1998) *Modeling the Time to Corrosion Initiation for Concretes with Mineral Admixtures and/or Corrosion Inhibitors in Chloride-Laden Environments*, PhD dissertation, Virginia Polytechnic Institute, Blacksburg, VA.

Zhang, D.; Gao, L.; Zhou, G. (2005) *Applied Surface Science*, 252, 4975.

Zhang, Y.; Yan, C.; Wang, F.; Li, W. (2005) *Corrosion Science*, 47, 2816.

Zhongping, X.; Wenli, H.; Wang, X.; Aigui, L. (2008) "Study on a Weatherable Coating System Which Could Be Applied on High Temperature (70°C) Steel Surface," *Proceedings*, 11th International Corrosion Symposium, Izmir, Turkey, 46–56.

Zinkevich, V.; Bogdarina, I.; Kang, H.; Hill, M. A. W.; Tapper, R. Beech, I. B. (1996) *International Biodeterioration and Biodegradation*, 37, 163–172.

# Index

# Also of Interest

## Check out these other related titles from Scrivener Publishing

### From the Same Author

*Corrosion Chemistry*, by Volkan Cicek and Bayan Al-Numan, ISBN 9780470943076. The causes and results of corrosion in industrial settings are some of the most important and difficult problems that engineers and scientists face on a daily basis. Coming up with solutions, or not, is often the difference between success and failure, and can have severe consequences. This timely volume covers the state-of-the art in corrosion chemistry today, for use in industrial applications and as a textbook. *NOW AVAILABLE!*

### Other Related Titles from Scrivener Publishing

*Bioremediation of Petroleum and Petroleum Products*, by James Speight and Karuna Arjoon, ISBN 9780470938492. With petroleum-related spills, explosions, and health issues in the headlines almost every day, the issue of remediation of petroleum and petroleum products is taking on increasing importance, for the survival of our environment, our planet, and our future. This book is the first of its kind to explore this difficult issue from an engineering and scientific point of view and offer solutions and reasonable courses of action. *NOW AVAILABLE!*

*An Introduction to Petroleum Technology, Economics, and Politics*, by James Speight, ISBN 9781118012994. The perfect primer for anyone wishing to learn about the petroleum industry, for the layperson or the engineer. *NOW AVAILABLE!*

*Ethics in Engineering*, by James Speight and Russell Foote, ISBN 9780470626023. Covers the most thought-provoking ethical questions in engineering. *NOW AVAILABLE!*

*Troubleshooting Vacuum Systems: Steam Turbine Surface Condensersand Refinery Vacuum Towers*, by Norman P. Lieberman. ISBN 978-1-118-29034-7. Vacuum towers and condensing steam turbines require effective vacuum systems for efficient operation. Norm Lieberman's text describes in easy to understand language, without reference to complex mathematics, how vacuum systems work, what can go wrong, and how to make field observations to pinpoint the particular malfunction causing poor vacuum. *NOW AVAILABLE!*

*Fundamentals of the Petrophysics of Oil and Gas Reservoirs*, by Buryakovsky, Chilingar, Rieke, and Shin. ISBN 9781118344477. The most comprehensive book ever written on the basics of petrophysics for oil and gas reservoirs. *NOW AVAILABLE!*

*Petroleum Accumulation Zones on Continental Margins*, by Grigorenko, Chilingar, Sobolev, Andiyeva, and Zhukova. Coming in September 2012, ISBN 9781118385074. Some of the best-known petroleum engineers in the world have come together to produce one of the first comprehensive publications on the detailed (zonal) forecast of offshore petroleum potential, a must-have for any petroleum engineer or engineering student. *NOW AVAILABLE!*

*Mechanics of Fluid Flow*, by Basniev, Dmitriev, and Chilingar. Coming in September 2012, ISBN 9781118385067. The mechanics of fluid flow is one of the most important fundamental engineering disciplines explaining both natural phenomena and human-induced processes. A group of some of the best-known petroleum engineers in the world give a thorough understanding of this important discipline, central to the operations of the oil and gas industry.

*Zero-Waste Engineering*, by Rafiqul Islam, ISBN 9780470626047. In this controversial new volume, the author explores the question of zero-waste engineering and how it can be done, efficiently and profitably. *NOW AVAILABLE!*

*Sustainable Energy Pricing*, by Gary Zatzman, ISBN 9780470901632. In this controversial new volume, the author explores a new science of energy pricing and how it can be done in a way that is sustainable for the world's economy and environment. *NOW AVAILABLE!*

*Formulas and Calculations for Drilling Engineers*, by Robello Samuel, ISBN 9780470625996. The most comprehensive coverage of solutions for daily drilling problems ever published. *NOW AVAILABLE!*

*Emergency Response Management for Offshore Oil Spills*, by Nicholas P. Cheremisinoff, PhD, and Anton Davletshin, ISBN 9780470927120. The first book to examine the Deepwater Horizon disaster and offer processes for safety and environmental protection. *NOW AVAILABLE!*

*Advanced Petroleum Reservoir Simulation*, by M.R. Islam, S.H. Mousavizadegan, Shabbir Mustafiz, and Jamal H. Abou-Kassem, ISBN 9780470625811. The state of the art in petroleum reservoir simulation. *NOW AVAILABLE!*

*Energy Storage: A New Approach*, by Ralph Zito, ISBN 9780470625910. Exploring the potential of reversible concentrations cells, the author of this groundbreaking volume reveals new technologies to solve the global crisis of energy storage. *NOW AVAILABLE!*